Studies in Big Data

Volume 25

Series editor

Janusz Kacprzyk, Polish Academy of Sciences, Warsaw, Poland
e-mail: kacprzyk@ibspan.waw.pl

About this Series

The series "Studies in Big Data" (SBD) publishes new developments and advances in the various areas of Big Data- quickly and with a high quality. The intent is to cover the theory, research, development, and applications of Big Data, as embedded in the fields of engineering, computer science, physics, economics and life sciences. The books of the series refer to the analysis and understanding of large, complex, and/or distributed data sets generated from recent digital sources coming from sensors or other physical instruments as well as simulations, crowd sourcing, social networks or other internet transactions, such as emails or video click streams and other. The series contains monographs, lecture notes and edited volumes in Big Data spanning the areas of computational intelligence incl. neural networks, evolutionary computation, soft computing, fuzzy systems, as well as artificial intelligence, data mining, modern statistics and Operations research, as well as self-organizing systems. Of particular value to both the contributors and the readership are the short publication timeframe and the world-wide distribution, which enable both wide and rapid dissemination of research output.

More information about this series at http://www.springer.com/series/11970

D.P. Acharjya · M. Kalaiselvi Geetha
Editors

Internet of Things: Novel Advances and Envisioned Applications

 Springer

Editors
D.P. Acharjya
School of Computer Science
and Engineering
VIT University
Vellore
India

M. Kalaiselvi Geetha
Department of Computer Science
and Engineering
Faculty of Engineering and Technology
Annamalai University
Chidambaram
India

ISSN 2197-6503   ISSN 2197-6511   (electronic)
Studies in Big Data
ISBN 978-3-319-85161-7   ISBN 978-3-319-53472-5   (eBook)
DOI 10.1007/978-3-319-53472-5

This Springer imprint is published by Springer Nature
The registered company is Springer International Publishing AG
The registered company address is: Gewerbestrasse 11, 6330 Cham, Switzerland

To
My wife Asima Nanda, and loving children
Aditi and Aditya

D.P. Acharjya

My husband B. Sivaraman, and loving
children Abhinaya and Khiran

M. Kalaiselvi Geetha

Preface

The Internet has restructured the global interrelations, the art of businesses, the cultural revolutions, and an unbelievable number of personal characteristics. Currently, machines are getting in to control innumerable autonomous gadgets via Internet and create Internet of Things (IoT). Thus, appliances are becoming the user of the Internet, just like humans with the Web browsers. Internet of Things is attracting the attention of recent researchers for its most promising opportunities and challenges. It has an imperative economic and societal impact for the future construction of information, network, and communication technology. The new regulation of future will be eventually, everything will be connected, and intelligently controlled. The concept of IoT is becoming more pertinent to the realistic world due to the development of mobile devices, embedded and ubiquitous communication technologies, cloud computing, and data analytics. Business procedures can be authorized; industries can be redesigned along IoT paradigm. In a broader sense, just like the Internet, Internet of Things enables the devices to exist in a myriad of places and facilitates applications ranging from trivial to the crucial. Conversely, it is still mystifying to understand IoT well, including definitions, content, and differences from other similar concepts.

The objective of this edited book was to provide the researchers of computer science and information technology the concepts, architectures, models, and key technologies which are required to achieve an in-depth knowledge in IoT. To achieve these objectives, we emphasized on essential concepts and its applications to real-life problems. This has been done to make the edited book more flexible and to stimulate further research interest in topics. We believe that our effort can make this collection interesting and highly attract the students pursuing pre-research, research, and even masters. This book is comprised of four parts: The first part is an attempt to provide an insight into theoretical foundation and fundamentals of on big data analysis that includes scalable architecture for big data processing, time series forecasting for big data, hybrid intelligent techniques, and applications to decision making by using neutrosophic sets. The second part discusses architecture for big data analysis and its applications, whereas the final parts discuss the issues pertaining to cloud computing.

Key deployment features in LTE-A systems for better support to IoT applications is explained in Chapter "Relay Technology for 5G Networks and IoT Applications." These features are likely to be improved further in 5G networks and future wireless mobile technologies. The exponential growth of wireless services driven by mobile Internet and smart devices has triggered the investigation of the 5G cellular network. For the better support of IoT applications, five key deployment features in LTE-A systems were discussed. Multi-hop relay provides coverage to mountainous and sparsely populated areas, heterogeneous relay saves bandwidth in the access link, mobile relay enhances the data rate of high-speed mobile users, and multiple backhauling provides higher peak data rates, and relay-assisted D2D enhances the average data rate. These features are likely to be enhanced further in 5G network for future wireless mobile technologies.

Two-way authentication security scheme for IoT based on existing Internet standards, specifically the datagram transport layer security protocol, is introduced and thoroughly discussed in Chapter "Two-Way Authentication for the Internet-of-Things." By relying on an established standard, existing implementations, engineering techniques, and security infrastructure can be reused, which enables an easy security uptake. The proposed security scheme uses two public key cryptography algorithms tailored for the resource heterogeneous nature of IoT devices. The extensive evaluation, based on real IoT systems, shows that the proposed architecture provides message integrity, confidentiality, and authenticity with affordable energy, end-to-end latency, and memory overhead.

Service-oriented architecture is essential for the use of wireless sensors networks in industrial applications such as the operation and maintenance of industrial installations. This architecture comprises the OCARI wireless sensor network and the OPC-UA/ROSA middleware, as well as the KASEM predictive maintenance system. This architecture targets various industrial applications such as process monitoring, pollutant detection, monitoring of fuel storage area, fire detection in temporary worksites, and health monitoring of people working in hazardous conditions. The solution proposed to support mobility in the OCARI network is simple and limits the overhead induced by mobile nodes. This mobility support is designed to be efficient in its use of resources. The properties of energy efficiency, determinism, latency, and robustness provided by OCARI to static wireless sensor nodes are ensured. In the absence of mobile nodes, the OCARI network behaves exactly as without mobility support and exhibits exactly the same performances. The design of such protocol extensions is also highlighted in Chapter "Mobility Support and Service Discovery for Industrial Process Monitoring."

Body area networks (BAN) represents the natural union between connectivity and miniaturization. Formally, it is defined as a system of devices in close proximity to a human body that cooperate for the benefit of the user. These networks are appealing to the researchers due to their wide range of applications. However, typical properties of body area network bring the necessity to achieve an efficient medium access protocol in terms of power consumption and delay. The strength of a BAN signal is affected by the physical location (in or on body) and orientation of the nodes, in relation to each other as well as the human body. A complete survey is presented in Chapter "MAC Protocols in Body Area Network-A Survey."

Current developments are observed in Internet of Nano-Things (IoNT) and Industrial Internet of Things (IIoT). Advanced development in nanotechnology led to nanomachines and nanoscale devices. The number of connected devices will increase at a hasty pace in future. This may drive added intensification in the network size and complexity for real-time traffic handling, since these devices are smart with manifold features. Adoption of IoNT will facilitate communication over Internet to facilitate interaction among these real-world physical elements. IIOT is heralded habitually as a way to improve operational efficiency. Chapter "Internet of Nano Things and Industrial Internet of Things" focuses on the concepts, architecture, applications, and future research directions in the both IoNT and IIoT.

Security issues in the deployment of wireless sensor networks in IoT are discussed in Chapter "Secure Distributed Group Rekeying Scheme for Cluster Based Wireless Sensor Networks Using Multilevel Encryption." A model that deals with a variety of passive attacks including node capturing and eavesdropping followed by a novel scheme using dynamic key management and encrypted data security is elucidated. With the evolution of diverse tiny devices, unrealistic connection among themselves and other devices has become practical. Most significant aspect in the IoT paradigm is WSN. Connecting WSN and other IoT elements go beyond remote access and create a heterogeneous information system. Embedding the computational competence in all objects present in the system provide a qualitative and quantitative leap in primary sectors such as health care and logistics. Enhanced usage of wireless sensor networks in fields such as military obligates the building up of secure environment. To this end, Chapter "Secure Distributed Group Rekeying Scheme for Cluster Based Wireless Sensor Networks using Multilevel Encryption" presents a novel scheme using dynamic key management and encrypted data protection to ensure environmental security with reduced and limited sensor usage.

Presently, wireless sensor networks are becoming more spreading and dominating. Both industry and academia are targeting their research works for the sake of advancing their functions. The safety of a wireless sensor network is negotiated due to the random distribution of sensor nodes in exposed environment, memory restraints, power restraints, and unattended nature. Furthermore, providing confidence between every couple of communicating nodes is a demanding issue in this kind of networks. Under these conditions, Chapter "Recognizing Attacks in Wireless Sensor Network in View of Internet of Things" spotlights on recognizing a variety of attacks and their symptoms on wireless sensor networks.

Association of wireless sensor network in IoT automation to enable green computing is of recent research. It combines sensing, computation, and communication into a single miniature device, thus necessitating IoT on wrathful utilization of WSN. Simultaneously, it does not assume a specific communication technology. But wireless communication technology plays a major role and in particular propagates many applications in many industries. The tiny, craggy, economical, and low-powered wireless sensor networks sensors bring the IoT to even the smallest objects installed in any environment at reasonable costs. Chapter "Wireless Sensor Network in Automation and Inter of Things" provides a survey on wireless sensor network in automaton.

Chapter "Challenges of Distributed Storage Systems in Internet of Things" describes the issues related to fusion, privacy, security, and trust in distributed storage systems to gain more benefits from IoT. Important information from the network devices in various places has to be sensed, accessed, and processed for various purposes. Additionally, the heterogeneous information acquired from diverse devices cannot be handled by traditional storage system. It makes distributed data storage more popular for proficient data management and leads to technologies such as cloud computing. The augmentation of these storage devices entails several challenges such as data consistency, error handling, data fusion, security, and privacy.

Chapter "Internet of Things Based Intelligent Elderly Care System" illustrates an approach for intelligent elderly care system on vision-based IoT. It explains a method for fall detection of elderly people which could prevent fatality and could provide immediate attention to other health-related injuries of vulnerable. A foremost challenge for society in near future will be meeting the needs of an aging people. Energy efficiency programs such as green deal miss many elderly in fuel poverty. Such people and those with health-related issues may be benefited by simply installing such an intelligent vision-based system in their homes. Moreover, the disadvantage of carrying a sensor by the sick and elderly is also prevented by identifying and sending alerts to other family members on acute emergencies.

The characteristics and applications of domain-specific IoTs, including smart cities, smart medical and health care, retail, logistics, supply chain management, manufacturing, aerospace and aviation, automotive and telecommunication, smart energy, smart transportation, smart pharmaceutical industry, and smart environment, are elucidated in Chapter "Challenges, Issues and Applications of Internet of Things." Also, IoT devices congregate and distribute information directly with each other and the cloud, making it achievable to gather, record, and analyze new data streams faster and more precisely. This chapter explores in greater depth the role of IoT in various applications with a quicker look at the technological characteristics that formulate it a reality and examine the opportunities, challenges, and issues at present.

The omnipresent exploitation of mobile and sensor devices is creating IoT a broad collection of potential Internet applications. It is apparent that the current hype around the IoT is enormous. The various information and communication perspective challenges associated with IoT for the expansion of the future society is stressed in Chapter "Application of Technologies in Internet of Things." Additionally, it also discusses how this technology can help in medical applications to offer cost-effective quality health care with enhanced manpower competence. With the maturity of the technology for collecting, analyzing, and transmitting data in the IoT, exciting novel IoT-driven healthcare applications and systems emerge progressively.

Current state of the Internet of Things from people's association point of view is much important. In IoT, objects become smart and autonomously communicate with one another and human beings, through networks supported by interfaces. The IoT systems are enhanced by surveying diverse interactions between the humans and the IoT to mine the implanted intelligence about individual, environment, and society. In the upcoming years, the IoT is expected to bridge various technologies to enable new applications by connecting physical objects together in support of intelligent decision making. With this perception, Chapter "An Appraisal on Human-Centered Internet of Things" spotlights and surveys the support of intelligent human–computer interaction for the IoT and to deal with human-centered concerns.

The Internet of Things represents the upcoming huge step in the Internet with its ability to gather, distribute, analyze, and interpret data. Millions of devices are expected to be connected or networked into the IoT structure that require massive dissemination of networks as well as the method of converting raw data into meaningful interpretations. The form of communication that is experienced now is either human–human or human–device. More influential smart phones, appliances, tablets, and the applications that are similarly rich and powerful available for each will enable buyers and business customers to interact seamlessly with companies altering the business processes. The technologies that enable the implementation of IoT, the objectives, future vision, and case studies are presented in Chapter "A Survey on Internet of Things: Case Studies, Applications, and Future Directions."

Internet of things and cloud computing are dissimilar technologies, but they are already part of our life. The espousal is considered to be more and more persistent and consequently making them as significant components of the future Internet. A novel paradigm of merging cloud computing and IoT is foreseen as disruptive which may enable a huge application scenario. Cloud computing is considered as a dynamic infrastructure which will be the only choice to maintain enormous and volatile information and can supply an illusion of unlimited computing resources to the users. Cloud will empower IoT by presenting resilient computing power, storage, and networking. With this perception, Chapter "Internet of Things in Cloud Computing" discusses cloud computing in IoT.

Much of researchers in different industries and organizations across the globe have started research in Internet of Things. Simultaneously, a lot of techniques pertaining to computational intelligence, data analysis, and cloud computing are progressing at the other end. Fusing the IoT and these latest techniques, technologies will acquire it to a newer dimension. To keep abreast with this development in a cohesive manner, we strove to keep the book reader friendly. The main objective is to bring most of the major developments in the above-mentioned area in a precise manner, so that it can serve as a handbook for many researchers. We trust and hope that this edited book will help the researchers, who have interest in IoT, cloud computing, and its applications to keep insight into recent advances and their importance in real-life applications.

Vellore, India   Dr. D.P. Acharjya
Chidambaram, India   Dr. M. Kalaiselvi Geetha

Acknowledgements

It is with great sense of satisfaction that we present our edited book and wish to express our views to all those who helped us both direct and indirect way to complete this project. First and foremost, we praise and wholeheartedly thank the almighty God, which has been unfailing source of strength, comfort, and inspiration in the completion of this project.

While writing, contributors have referred to several books and journals, and we take this opportunity to thank all those authors and publishers. We are extremely thankful to the reviewers for their constant support during the process of evaluation. Special mention should be made of the timely help given by different persons during the project work, those whose names are not mentioned here. Last but not the least, we thank the series editor "Janusz Kacprzyk" and the production team of **Springer** for encouraging us and extending their cooperation and help for a timely completion of this edited book. We trust and hope that it will be appreciated by many readers.

Contents

Contributors

D.P. Acharjya VIT University, Vellore, Tamilnadu, India

Nasim Afshari Department of Knowledge Engineering and Decision Sciences, Kharazmi University, Tehran, Iran

N. Syed Siraj Ahmed VIT University, Vellore, Tamilnadu, India

J. Arunnehru Department of CSE, Annamalai University, Chidambaram, India

Patrick Bellot LTCI, Télécom ParisTech, Université Paris-Saclay, Paris, France

Abderrahmane BenMimoune Department of Electrical Engineering, ETS-Quebec University, Montreal, QC, Canada

Tuan Dang EDF R&D Lab, STEP Department, Chatou Cedex, France

Mayank Dave National Institute of Technology, Kurukshetra, India

Omid Mahdi Ebadati E. Department of Mathematics & Computer Science, Kharazmi University, Tehran, Iran

A. Geetha Department of Computer Science and Engineering, Annamalai University, Tamilnadu, Chidambaram, India

P. Geethanjali School of Electrical Engineering, VIT University, Vellore, Tamilnadu, India

Hemdan Ezz El-Din Department of Computer Science, Mangalore University, Mangalore, India

Wen Hu School of Computer Science and Engineering, The University of New South Wales, Sydney, NSW, Australia

Michel Kadoch Department of Electrical Engineering, ETS- Quebec University, Montreal, QC, Canada

M. Kalaiselvi Geetha Department of Computer Science and Engineering, Annamalai University, Tamilnadu, Chidambaram, India

Thomas Kothmayr Fakultät Für Informatik, Technische Universität München, Garching, Germany

Manish Kumar PEC University of Technology, Chandigarh, India

Erwan Livolant Inria, Paris, France

D.H. Manjaiah Department of Computer Science, Mangalore University, Mangalore, India

Murali Manohar VIT Business School, VIT University, Vellore, Tamilnadu, India

Pascale Minet Inria, Paris, France

Christophe Mozzati Predict, Vandoeuvre-les-nancy, France

Faezeh Nayyeri Department of Knowledge Engineering and Decision Sciences, Kharazmi University, Tehran, Iran

Kalavathy Perumal VIT Business School, VIT University, Vellore, Tamilnadu, India

A. Punitha Department of CSE, Annamalai University, Annamalainagar, Tamil Nadu, India

G.R. Sakthidharan Department of CSE, Gokaraju Rangaraju Institute of Engineering and Technology, Hyderabad, Telangana, India

Corinna Schmitt Communication Systems Group CSG, Department of Informatics IfI, University of Zurich, Zurich, Switzerland

Madhvaraj M. Shetty Department of Computer Science, Mangalore University, Mangalore, India

Burkhard Stiller Communication Systems Group CSG, Department of Informatics IfI, University of Zurich, Zurich, Switzerland

Mohsen Yazdinejad Department of Knowledge Engineering and Decision Sciences, Kharazmi University, Tehran, Iran

Acronyms

AEAD	Authenticated Encryption Associated Data
AFS	Andrew File System
AODV	Adhoc On demand Distance Vector
AP	Access Point
ARP	Address Resolution Protocol
AT	Assistive Technology
ATM	Asynchronous Transfer Mode
ATT	Air Touch Technology
BAN	Body Area Network
BFFS	Berkeley Fast File System
BLIP	Berkeley Low-Power Internet Protocol
CA	Collision Avoidance
CBC	Cipher Block Chaining
CCA	Clear Channel Assessment
CCAP	Configurable Contention Access Period
CCU	Central Control Unit
CEP	Complex Event Processing
CFP	Contention Free Period
CMMS	Computerized Maintenance Management System
CoAP	Constrained Application Protocol
CORBA	Common Object Request Broker Architecture
CPAN	Coordinator of Personal Area Network
CSMA	Carrier Sense Multiple Access
DCOM	Distributed Component Object Model
DDS	Data Distribution Service
DHT	Distributed Hash Table
DLL	Dynamic Link Library
DoS	Denial of Service
DSRC	Dedicated Short Range Communication
DTLS	Datagram Transport Layer Security

ECC	Elliptic Curve Cryptography
ECDH	Elliptic Curve Diffie–Hellman
ECDSA	Elliptic Curve Digital Signature Algorithm
EDCA	Enhanced Distributed Channel Access
EMC	Electromagnetic Compatibility
EOFS	European Open File System
EOLSR	Energy-Efficient OLSR
EPAIFS	Emergency Priority Arbitrary Inter-Frame Space
EPC	Electronic Product Code
EPSIFS	Emergency Priority SIFS
ETSI	European Telecommunications Standards Institute
FDD	Frequency Division Duplex
FDDI	Fiber-Distributed Data Interface
FTP	File Transfer Protocol
GFS	Google File System
GPFS	General Parallel File System
GPP	Generation Partnership Project
GPRS	General Packet Radio Service
GPS	Global Positioning System
HAN	Home Area Network
HBC	Human Body Communication
HCI	Human–Computer Interaction
HCS	Human-Centered Systems
HDFS	Hadoop Distributed File System
HMAC	Hash-Based Message Authentication Code
HMI	Human-to-Machine Interfaces
HMM	Hidden Markov Model
HTTP	Hypertext Transfer Protocol
HTTPS	Secured Hypertext Transfer Protocol
HWSN	Heterogeneous WSN
ICMP	Internet Control Message Protocol
IGRP	Interior Gateway Routing Protocol
IIoT	Industrial Internet of Things
IKE	Internet Key Exchange
IoBNT	Internet of Bio–Nano-Things
IoMNT	Internet of Multimedia Nano-Things
IoNT	Internet of Nano-Things
IoT	Internet of Things
IV	Initialization Vector
KASEM	Knowledge and Advanced Services for E-Maintenance
KNN	K-Nearest Neighbor
LAN	Local Area Network
LEE	Light-Weight Energy-Efficient Encryption
LGPL	Lesser General Public License
LPL	Low-Power Listening

LTE	Long-Term Evolution
MAC	Media Access Control
MANET	Mobile Adhoc Network
MEMS	Micro-Electro Mechanical Sensors
MPP	Motion Projection Profile
MQTT	Message Queuing Telemetry Transport
MTU	Maximum Transmission Unit
NACK	Negative Acknowledgment
NAS	Network-Attached Storage
NAV	Network Allocation Vector
NFC	Near-Frequency Communication
NPSIFS	Normal Priority SIFS
NSD	Network Shared Disk
OCARI	Open Communication Protocol for Adhoc Reliable Industrial Instrumentation
OMG	Object Management Group
ON	Overlay Network
ONS	Object Name Server
OPC-UA	Open Platform Communications-Unified Architecture
OSERENA	Optimized Schedule Router Nodes Activity
OSPF	Open Shortest Path First Protocol
PAN	Personal Area Network
PCB	Printed Circuit Board
PHR	Personal Health Record
PIKE	Peer Intermediaries for Key Establishment
PKC	Public Key Cryptography
PLR	Packet Loss Rate
PMP	Point to Multipoint
PRB	Physical Resource Block
QoS	Quality of Service
RAID	Redundant Array of Independent Disks
RAN	Radio Access Network
RARP	Reverse Address Resolution Protocol
RBF	Radial Basis Function
RDP	Remote Desktop Protocol
RF	Radio Frequency
RFC	Remote Function Call
RFID	Radio Frequency Identification
RIP	Routing Information Protocol
ROI	Region of Interest
ROSA	Robust Overlay Network with Self Adaptive Topology
RPL	Routing Protocol for Low power and Lossy Networks
RSSI	Received Signal Strength Indication
RTOS	Real-Time Operating System
RTT	Round Trip Time

SAN	Storage Area Network
SASPKC	Stateful Public Key Cryptography
SCADA	Supervisory Control and Data Acquisition
SIFS	Short Inter-Frame Space
SiP	System in Package
SMTP	Simple Mail Transfer Protocol
SOA	Service-Oriented Architecture
SoC	System on Chips
SSL	Secure Service Layer
SSL	Secure Sockets Layer
SVDD	Support Vector Data Description
SVM	Support Vector Machine
TaMAC	Traffic Adaptive MAC
TCP	Transmission Control Protocol
TDD	Time Division Duplex
TDMA	Time Division Multiple Access
TPM	Trusted Platform Module
TRL	Technology Readiness Level
UAKAS	User Authentication and Key Agreement Scheme
UDP	User Datagram Protocol
URL	Uniform Resource Locator
URN	Uniform Resource Network
UUID	Universal Unique Identifier
UWB	Ultra-Wideband
VPN	Virtual Private Network
WISP	Wireless Identification and Sensing Platform
WPA	Wireless Protected Access
WPAN	Wireless Personal Area Network
WSN	Wireless Sensor Networks
XML	Extensible Markup Language

Part I
Fundamentals and Architecture
of Internet of Things

Relay Technology for 5G Networks and IoT Applications

Abderrahmane BenMimoune and Michel Kadoch

Abstract Relaying technologies have been actively studied in mobile broadband communication systems, and were considered in the most recent standard releases of the Third Generation Partnership Project (3GPP), including "Long Term Evolution Advanced" (LTE-A) networks. This chapter provides an in-depth review of the relay technology that is being considered for future 5G networks. The article first introduces and compares different relay types that use LTE-A standards, and presents the relay benefits in terms of performance and operational costs. It then highlights future relay deployment strategies that have been discussed by the 3GPP, which supports multi-hopping, mobility, and heterogeneity. In addition, it also proposes efficient deployment strategies, along with their impact on network performance. Finally, the chapter explains a few of the associated challenges that lie ahead for relay application, and provides a video streaming application.

1 Introduction

The conventional topology of current cellular networks is a star structure, where central control points usually serve as base stations. This provides the advantage of simplicity of the architecture while still providing quality of service (QoS) guarantees. However, for next-generation networks, this topology will be disadvantageous and difficult to use due to the insufficient availability of network access. This high dependency on the central node has its own drawbacks (including inefficient offload), since all user data needs to go through the central node network. This topology also cannot offer performance and energy efficiency for users on the edge of cell coverage (edge users). Thus, the next-generation cellular network topology needs to be optimized, and relay will play an important role in this.

A. BenMimoune (✉) · M. Kadoch
Department of Electrical Engineering, ETS- Quebec University, Montreal, QC, Canada
e-mail: A.B.BenMimoune@ieee.org

M. Kadoch
e-mail: Michel.Kadoch@etsmtl.ca

© Springer International Publishing AG 2017
D.P. Acharjya and M. Kalaiselvi Geetha (eds.), *Internet of Things: Novel Advances and Envisioned Applications*, Studies in Big Data 25, DOI 10.1007/978-3-319-53472-5_1

In the next-generation cellular network, cellular networks will be able to include different kinds of relays. Due to the use of relaying technologies into the centrally controlled star network, different kinds of connections such as user equipment (UE) to UE, relay station (RS) to RS, and base station (BS) to RScan thus be established. The hybrid topology radio network will thus naturally be the future mobile access network, which can help to overcome current and future difficulties and challenges in an efficient manner.

Relay technology is also promising in Internet of Things (IoT) applications. In IoT, relaying can have many more functions in a cellular network, such as improving the topology of the cellular network, improving the robustness of a network, and decreasing power consumption. In addition, a multi-hop topology can efficiently support tremendous access for fog and social networking services.

The capacity offered by a macro cell is not uniformly distributed across its coverage areas in today's LTE network. From a users perspective, proximity to the cell center thus should result in much higher throughput and larger battery savings (due to reduced transmission power) than cell edge users. The 3GPP has standardized and begun supporting LTE relays in Release 10, with no impact on UE design and implementation. Relay stations are defined as low-power nodes that can be deployed underneath macro BSs, forming small cells to address the need for coverage and capacity improvements. One of the most attractive features of 3GPP relays is the LTE-A-based wireless backhaul (i.e., self-backhauling), as this can provide a simple deployment technique to improve coverage to dead zones (e.g., at cell edges) and, more importantly, traffic hot zones [1]. Communication between BSs and RSs is similar to the conventional communication between BSs and UEs, in which communication is accomplished via a point-to-multipoint (PMP) connection; this creates multiple paths from a BS to different RSs, and can also establish PMP connectivity with the UEs. In the following, the user connected to RS are denoted by R-UE and those connected directly to Macro BS by M-UE [2].

2   Relay Classification

Relays can be classified based on several characteristics, such as operation layers, duplexing, and the resources assignment scheme, all described below.

2.1   Classification Based on Layers

In this form, relay classification is based on the layers of the protocol stack in which their main functionality is performed. There are three main varieties:

1. A layer 1 relay, also called a booster or repeater, takes the received signal, amplifies it, and forwards it to the next hop, which may be another relay or UE. As its name implies, it works only at the physical layer. Although the advantage of this type is that they are inexpensive and simple, with minimum impact on the existing standard, they amplify noise simultaneously with the desired signals.
2. A layer 2 relay, called a decode and forward relay, can eliminate noise signals and perform radio resource management functions better than a layer 1 relay; a better throughput-enhancement effect thus can be expected. However, due to the extra functions performed by a layer 2 relay, a significant processing delay is introduced, due to modulation/demodulation and encoding/decoding.
3. A layer 3 relay, also called a decode and forward relay, can be thought of as a BS that uses a wireless rather than wired link for backhaul. A layer 3 relay can improve throughput by eliminating inter-cell interference and noise. However, This type of relay has an impact on standard specifications, in addition to the delay caused by modulation/demodulation and encoding/decoding processing.

2.2 Classification Based on Duplexing Schemes

In general, a relay station communicates through two links: with its donor base station on one hand (through the backhaul link) and its connected users on the other (through the access link). The relay station can use either a time division duplex (TDD) or a frequency division duplex (FDD) scheme in these communications.

In a basic TDD scheme, the downlink and uplink frames of the BSRS and RSUE links are usually not enabled simultaneously. For example, as shown in Fig. 1a, the two hop downlink transmissions occur in the first and second time slots, respectively, followed by the two hop uplink transmissions in the third and fourth time slots,

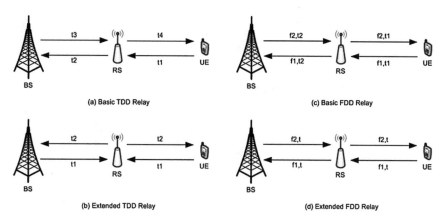

Fig. 1 TTD and FDD schemes

respectively. It can be seen that only one station can transmit and/or receive at each time slot. However, a better utilization can be achieved if more than one station can transmit and/or receive at each time slot. Hence, an extended TDD scheme can be used to improve resource efficiency, as shown in Fig. 1b.

In the standard FDD scheme, in contrast, the downlink and uplink transmission between the backhaul link will occur in the same time slot but at different frequencies; the same will occur for the access link case, as shown in Fig. 1c. However, for more efficiency, an extended scheme of the basic FDD approach can be designed to use

Fig. 2 In-band and out-band scenarios

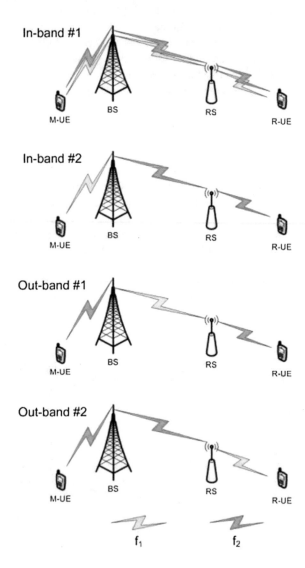

orthogonal frequencies for each link, which will allow the backhaul and access link transmissions to occur at the same time, as shown in Fig. 1d.

2.3 Classification Based on Resources Assignment

Relay nodes can be classified into inband and outband relay stations, according to the spectrum used for the backhaul link. Figure 2 shows the inband and outband scenarios for relay integration into the cellular radio access network (RAN). In the inband case, the backhaul link between the BS and RS shares the same carrier frequency with the RSUE links (scenario 1 share all the spectrum between different links, while scenario 2 dedicates a part of the spectrum to direct links and the rest for backhaul and access links). The backhaul link, in contrast, uses a separate frequency band for the outband case (scenario 1 share the spectrum between the direct and access links and uses a separate frequency band for the backhaul links, while in scenario 2 the separate frequency band is used by the access links).

2.4 3GPP Classification

The 3GPPs LTE-A standards have defined two types of RSs: type-1 and type-2 relay, also called non-transparent and transparent relay, respectively [3]. Relay classification is based on the relays ability to generate its own cell control message [4].

- Type-1 (non-transparent) relay is usually responsible for assisting UEs that are distant from the base station and out of cell coverage range. Thus, a type-1 relay will generate its own cell control messages to extend the signal and service coverage, while improving the overall system capacity. The type-1 relay has a two main extended modes:

  - Type-1a relay operating on outband mode with full duplex scheme.
  - Type-1b relay operating on inband mode with full duplex scheme.

- Type-2 (transparent) relay is responsible for helping UEs within the coverage area of the BS. Although the UE can communicate directly with the BS, the deployment of type-2 RS can help to improve its service quality and link capacity. Since this

Table 1 Summary of relay classifications

3GPP class	Layer	Integration into BS	Duplex scheme
Type 1	3	Inband	Half-duplex
Type 1a	3	Outband	Full-duplex
Type 1b	3	Inband	Full-duplex
Type 2	2	Inband	Full-duplex

type of relay is a part of the donor cell, cell control messages are not necessary in this case to improve the overall system.

Table 1 summarizes the different relay classifications discussed above, as well as their mapping.

3 Potential Benefits of Relaying

This section presents the potential benefits of employing relay stations in cellular networks. The following sections describe the main motivations for the use of relay stations.

3.1 Relay to Improve Channel Reliability

Cellular communication suffers from fading, path loss, and shadowing factors, which affect communication performance and tend to reduce its reliability. Relaying strategies in cellular networks can increase the reliability of the communications against these channel impairments by exploiting spatial diversity [5]. When a communication channel between the BS and a user is unreliable, relays can be used as repeaters to forward the data toward the user. As a result, the user will receive several copies of the transmitted signal over different transmission paths and can combine the data received to improve transmission reliability, as illustrated in Fig. 3.

Fig. 3 Spatial diversity for reliability improvement

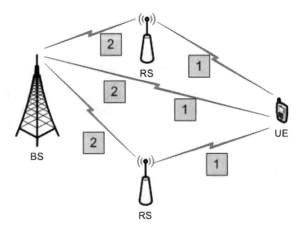

Fig. 4 Spatial diversity for resource aggregation

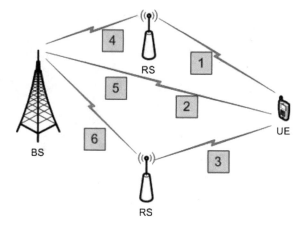

Fig. 5 Relay for coverage extension

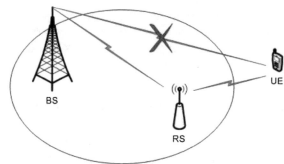

3.2 Relay to Improve System Throughput

Total system throughput can be increased when using relays by aggregating the resources offered from different stations, where the data blocks are transmitted from the BS along multi-paths toward the user [6]. Contrary to the previous scenario, where the same data are transmitted in different paths, when improving the system throughput, different transmission paths carry different data blocks. This has the effect of increasing the total transmission data rate between the base station and the user. As illustrated in Fig. 4, the user is connected to several stations with sufficient resources; as such, when these resources are aggregated, the throughput achieved by users can be increased.

3.3 Relay to Improve Service Continuity

Service continuity interruption can occur for different reasons, such as coverage holes. Relay can be used as a convenient method for filling coverage holes and

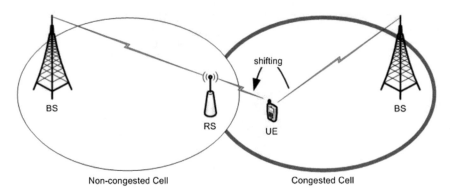

Fig. 6 Relay for traffic load-balance

extending coverage. In Fig. 5, when the service is interrupted between the base station and users, service still can be maintained using a relay path. In this case, a relay station will create a substitute path between the BS and users.

3.4 Relay for Offloading and Load Sharing

Several techniques for cellular traffic offloading have been proposed to improve the user experience in overloaded areas. Relaying technology is one technique that can be applied to balance traffic load among highly and lightly loaded cells. As shown in Fig. 6, preinstalled relay stations can be deployed to regulate traffic from highly to lightly loaded cells by shifting a set of users to a different donor base station. In this context, a relay station will load-balance traffic among macro cells by accepting a set of users within its coverage [7].

3.5 Relay to Reduce Operational Costs

Relaying in cellular networks can easily reduce operation costs for service providers. For example, the transmission power can be reduced significantly by deploying relay stations in the appropriate locations. This reduction can simply be due to the reduction of the path loss, which translates into reduced operational costs. Another example is coverage extension, where relay stations can provide an easy and cheap method of extending coverage without the need to install backhaul. Deployment costs and time thus are significantly reduced compared to traditional base stations [8].

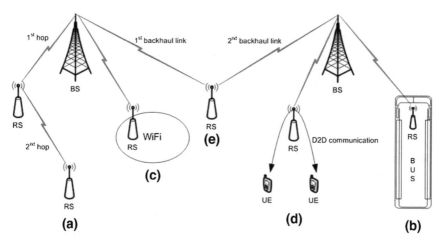

Fig. 7 Hybrid topology with relay in 5G networks

4 Relay Deployment Strategies in 5G Networks

The 3GPP has discussed various scenarios in which the introduction of relay technology will be potentially useful. The basic scenario for the use of relays can be expanded by factoring in the way in which relays can help to achieve the benefits mentioned previously. Fig. 7 shows the deployment scenarios. In scenario (a), multi-hop relay communication is an important scenario for operators to extend the coverage area to mountainous and sparsely populated regions. Case (b) shows the mobile relay scenario, in which relay stations are installed on vehicles (such as trains and buses) to improve throughput and reduce the volume of control signals from moving mobile stations. Scenario (c) shows a heterogeneous relay scenario, in which the relay uses the LTE network in backhaul links and the WLAN network for access links. In scenario (d), the relay station is used to relay a device-to-device (D2D) communication and offload the donor base station. Finally, in scenario (e), the relay is connected to several donor base stations to improve throughput and to balance loads across the base stations. In each of these scenarios, the type of relay that will be used may vary according to the specific requirements of each scenario.

4.1 Multi-hop Relay

It has been more than a decade since the multi-hop cellular network architecture was first proposed and analyzed [9]. Fundamental research projects have demonstrated the benefits of multi-hop architecture in terms of system capacity, service coverage, and network connectivity. The concept behind the multi-hop architecture could be considered to be a hybrid of mobile ad hoc networks (MANETs) and cellular net-

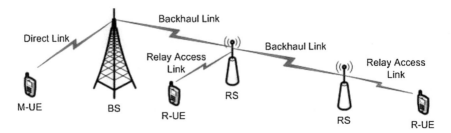

Fig. 8 Multi-hop relay deployment

works. Today, with the recent LTE-A standards, the 3GPP supports a single-hop relay technology, in which the relay station can be fixed or mobile and the radio access link between the BS and UE is relayed by only one relay station [10]. However, with the help of multi-hop relay, the radio link between the BS and UEs can be extended into more than two hops; the propagation conditions of each hop are expected to be better than the direct link used between the BS and UE in conventional cellular networks (as shown in Fig. 8).

Improving coverage and network capacity is the leading motivation for integrating multi-hop relays into LTE-A networks. This comes from the reduction of path loss due to the employment of multiple hops to transmit data to/from the corresponding base station. In [11], it was shown that better performance in terms of throughput, packet loss, and delay can be achieved by supporting multi-hop-relaying functionality. Using a multi-hop relay system, however, requires more radio resources to transmit data through different hops. More interference is also created due to a larger number of simultaneous transmissions in the network [12]. New mobility and resource allocation schemes thus are important for achieving a high QoS while increasing the whole network capacity.

4.2 Mobile Relay

Due to the high penetration of smartphones and tablets, the number of users who use wireless broadband services on public transportation is growing rapidly. The best solution to serve such users is to place a relay station as close as possible to the vehicle to compensate for the vehicular penetration loss. In practice, because the positions of vehicles are not known beforehand, the use of mobile relay is more economical and applicable to serve vehicular users, as shown in Fig. 9. In a recent 3GPP study, a mobile relay deployment scenario was considered to be a cost-effective solution to serve data-intensive users using public transportation [13]. In addition, the group handoff can be performed by regarding the users who are served by the same mobile relay as a group, which could reduce the probability of handoff failure. However, the backhaul link can be considered as the capacity bottleneck of the deployment,

Fig. 9 Mobile relay
deployment

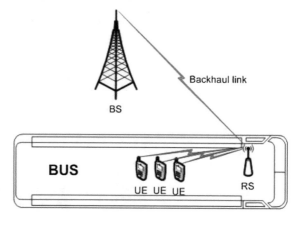

Fig. 10 Heterogeneous
relay deployment

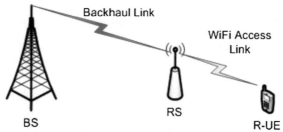

particularly in high-speed scenarios, where more resources need to be allocated
compared to fixed relay stations in a similar position.

4.3   Heterogeneous Relay

The WiFi access network is currently very popular, and most mobile devices and
laptops are WiFi-capable. Heterogeneous relay is attractive for covering a specific
local area with WiFi on the access link while using the LTE network on the backhaul
link, as shown in Fig. 10. Using heterogeneous relay combined with mobile relay
to provide WiFi Internet access to onboard data users is fairly common. The most
attractive quality when using WiFi air interface for the access link is having the
opportunity to serve all mobile terminals without subscribing to the operator owner
of the backhaul link; this leads to the ability to optimize the number of relays, instead
of having different relays for each operator. Although WiFi-only devices can also
use the cellular network, which may provide extra income to service providers, using
WiFi technology to provide UE with a seamless experience in the current cellular
network is challenging. In addition, because WiFi networks operate on the open
industrial, scientific, and medical radio (ISM) bands, any interference in these radio
bands cannot be coordinated in the same manner as in the dedicated frequency bands

Fig. 11 Relay-assisted D2D
deployment

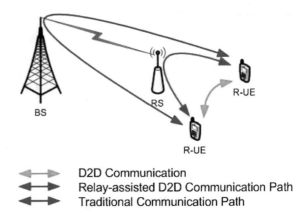

owned by operators. It is thus difficult for operators to offer similar QoS as in their
own cellular networks [14].

4.4   Relay Aassisted D2D Communication

In the context of D2D communication, it is crucial to set up reliable direct links
between the UEs while satisfying the QoS of traditional cellular and D2D users in
the network. The excessive interference and poor propagation channel may also limit
the advantages of D2D communication in practical scenarios. In such cases, with the
support of relaying technology, D2D traffic can be transmitted via relay, as shown in
Fig. 11. Relay thus can efficiently enhance the performance of D2D communication,
particularly when D2D pairs are too distant from each other, or the quality of the
D2D channel is not good enough for direct communication [15].

4.5   Multiple Backhaul Relay

When referring to relay architecture, the common scenario is that there is a point-
to-multipoint communication between RSs and BS, where multiple RSs can be connected to a BS, but an RS is connected to only one BS. Through such an architecture,
relay deployment can be straightforward and simple. However, this architecture might
limit the system performance because of the capacity available on the backhaul link,
particularly when the neighboring cells are lightly loaded [16]. Multiple backhaul
relay deployment can enable the many-to-many connections scheme between RSs
and BSs, where the RSs can be connected to multiple BSs through the UE interface,
as shown in Fig. 12. Although this relaying scenario of deployment is completely
transparent to the UEs and will make the system more flexible by creating alterna-

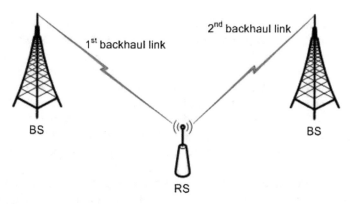

Fig. 12 Multiple backhaul relay deployment

tives for load sharing, changes will be required in the BS, RS, and the core network. The most significant change is the cooperation mechanism at the BS, RN, and core network, and the reordering of uplink data arriving via several S1 interfaces at the core network, and downlink data arriving via several UE interfaces at the relay station [16].

5 Future Challenges and Research Directions

While relaying within cellular networks has various potential benefits, several challenges do arise when relay stations are deployed. To support the integration of relays, a few adjustments and adaptations are required in the medium access control and network layers. In the following section, a discussion of several challenges to achieving the benefits discussed earlier, and the required modifications.

5.1 Radio Resource Allocation

In the LTE network, the radio resources are composed of physical resource blocks (PRB), which have both a time and frequency dimension. Thus, different users share the PRBs that can be assigned by the BS scheduler. In addition, in the presence of an RS, the radio resources at each station (BS and RSs) are shared between the direct, backhaul, and the access links. It is thus essential to design efficient resource allocation schemes in the presence of relay stations. A dynamic resource allocation scheme may also facilitate the scheduling algorithm and interference management [17].

One of the most important points in radio resource allocation is to determine whether to use a centralized or a distributed strategy. Within the context of conventional cellular networks, the resource allocation scheme is considered centralized to the base station. However, within the context of relaying, a resource allocation scheme can be centralized if the central node is the BS, and can be distributed if the resource allocation algorithm is implemented in the RSs. In centralized resource allocation schemes, the resource allocation algorithm is implemented in one central node (usually the BS); all data from relay stations is transmitted into this node to execute the allocation algorithm and to allocate resources. The problem is that the central node becomes a bottleneck of the network and needs to be very powerful. In distributed resource allocation schemes, in contrast, the allocation algorithm is implemented separately at the BS and the RSs, which results in the signaling overhead being substantially reduced. The problem in the distributed algorithms is that they are unable to achieve optimal resource allocation, as there is no central node with the ability to efficiently coordinate and control the resources that are used between dif-

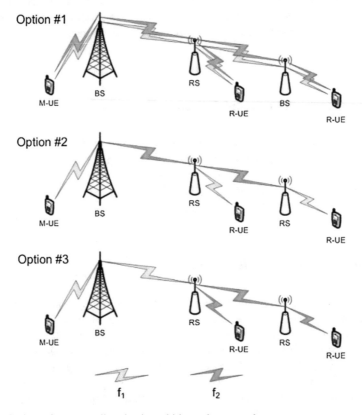

Fig. 13 Options of resource allocation in multi-hop relay networks

ferent links. A suboptimal distributed resource allocation algorithm could be adopted as a solution, with minimal overhead in multi-hop relay networks.

All the radio resources in the networks with layer 1 relay are centrally managed at the BS. Layer 2 and layer 3 relays have their own resource allocation functions, so that the UEs located within the relay coverage may be managed by the RS itself. Fig. 13 shows an example of resource allocation schemes in a multi-hop relay scenario. These examples present just an overview of the possible options, and is not an exhaustive list of FDD and TDD variations.

5.2 Power Control

The design of traditional cellular networks tends to maximize capacity and coverage, which can potentially lead to solutions where energy efficiency drops. In relaying networks, energy efficiency can be seen from two viewpoints. On the operator side, the energy spent by the infrastructure may increase by increasing the number of relay stations, implying high operational costs. On the user side, although the relays reduce the energy consumption by bringing the RAN closer to the end user, some communication strategies require a high computational burden at the UE side, which has a negative impact on battery lifetime [18].

The intelligent use of energy is thus vital in achieving efficient energy usage and interference coordination, particularly in relaying networks when considering all users (i.e., users connected directly and users connected through relays) to be equals and when their QoS requirements are delivered with the same priority. In addition, in conventional cellular networks, it was always assumed that the base station had free (meaning unlimited) access to energy; this is not necessarily true for relaying scenarios, particularly in mobile relay where the relay can have a limited power source. The power allocation scheme can also be centralized in the BS or distributed among the relays, which makes it more complex to employ. The power control mechanism can be considered jointly with the admission control mechanism and a resource allocation strategy to optimize the network performance. Power efficiency for relaying in cellular networks thus remains a challenge that needs to be solved.

5.3 Admission Control

In a conventional cellular network, the base station manages the admission control policy. In other words, if a user wishes to establish a new call and the radio resources are not available in its home BS, he or she will be automatically blocked. In the relaying context, the admission control should be coordinated between the BS and its subordinate relays, thus generating a lot of delay as well as signaling overhead (particularly for R-UEs), while the resource availability of the access links cannot be the only metric for decision making. The resource availability on the backhaul links also should be considered in order to guarantee the availability of resources through the wireless radio path. These relaying schemes thus need to be well designed and operating as intended in order to reduce call blocking probability and to improve QoS fairness in terms of call blocking probability by balancing traffic among congested and non-congested cells.

5.4 Relay Selection

The relay selection process specifies the user assignment scheme to the different stations that are available in the network. Usually in conventional cellular networks, the UE selects one of the base stations available based on the channel condition, such that the user will achieve a high level of throughput. In the relaying context, however, the relay selection problem can be more complex, since users may have several relay stations in their range that they can choose from: particularly since the user-achieved throughput is mainly affected by the decision to associate a user with one station. In addition, the conventional rule in a user association scheme is usually that a user can only be connected to one station at a time. However, in a dense relay deployment, a multiple relay selection can also be employed to enhance the user data rate.

The selection scheme can be executed in either a centralized or distributed manner. In a centralized scheme, each RS will collect channel and location information from users in its vicinity and will forward the information to the donor BS, which will serve as a central point for making appropriate pairing decisions. In a distributed selection scheme, in contrast, each RS selects an appropriate UE in its neighborhood based on local channel information. Generally speaking, centralized schemes require more signaling overhead, but can achieve better performance gains than their distributed counterparts.

5.5 Handoff

The handoff mechanism plays a critical role in the mobility management protocol design. In conventional cellular networks, the user only has the option to execute a handoff between BSs. In the relaying context, on the other hand, the handoff algorithm might be more complex, and could require the use of more coordination and cooperation mechanisms to provide a guaranteed QoS. The introduction of relay stations enables several handoff scenarios, which can be categorized into intra-BS and inter-BS handoff. The main difference in these scenarios is which nodes are the old access station and which are the new access station. If both the new and old access stations are in the same BS cell, then it is an intra-BS handoff procedure; otherwise, it is an inter-BS handoff, as shown in Fig. 14.

The increasing density of relays in cellular networks may also increase the handoff rate, since more cell boundaries will be present in the network, which provides more

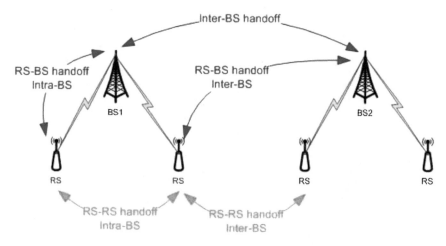

Fig. 14 Handoff scenarios in relaying networks

opportunities for the UE to be handed over. As such, the classical handoff metrics (such as location and signal strength) may not be enough in decision-making; other relaying metrics should also be taken into consideration, such as the load of the backhaul links and the target relay cell. For the mobile relay context, the handoff mechanism is applied to the UEs, as well as to the mobile RS itself. The handoff design issue is therefore common for users and mobile RS. The mobile RS handoff also needs to consider group mobility issues, as well as the subsequent configuration of the whole mobile group [19].

5.6 Routing

The traditional cellular network is based on a star topology, where all UEs are connected directly (single-hop) to the BS. The routing functionality therefore has no significance in this scenario. In the relaying context, however and particularly in multi-hop relay networks with multiple backhaul capabilities the purpose of a routing protocol is to find the appropriate radio path for each user to establish a connection to its home BS (Fig. 15). In cellular networks, each users access is typically based on the criterion of maximizing received signal strength. This routing criterion has not taken into account the characteristics of relay transmission, particularly the backhaul link. Since two-hop transmission may lead to increased delay, bandwidth utilization, and packet loss, it is more reasonable to use the QoS metrics as the routing criteria. A fundamental question that still needs to be solved in multi-hop relay networks is therefore how to perform joint relay selection and routing, such that maximum performance gains (in terms of network capacity, coverage, and QoS performance) can be achieved.

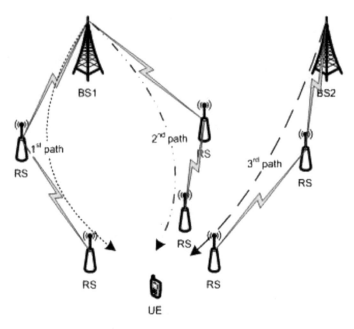

Fig. 15 Example of routing in multi-hop relay network's

6 Fog Computing and Video Streaming Applications

This section presents the benefits of employing a multi-hop relay topology to support the access for fog networking services. The following sections presents a video streaming application in a distributed architecture.

6.1 Fog Computing Paradigm

Both academic researchers and industry practitioners have expressed great interest in the cloud computing paradigm in recent years, since it offers ways to avoid the computation/storage limitation of UEs and provides elastic resources to the applications used on these devices. However, a few problems still remain unsolved in certain IoT applications that require high mobility support, geo-distribution, location-awareness, and low latency. Fog computing was proposed as an alternative solution to enable distributed computing at the edge of the network; this can deliver new applications and services, especially for future Internet development [20]. Relay stations provide potential infrastructures that can provide resources for services at the edge of the network; they can play the role of fog nodes by increasing their processing speed and adding a local network storage. The relaying infrastructure can also provide a better QoS for fog services by supporting the edge caching functionality, as illustrated

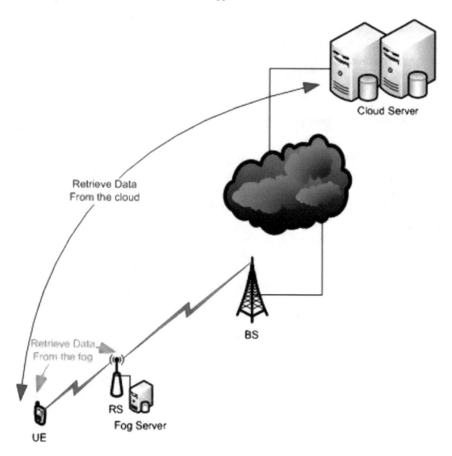

Fig. 16 Cloud and Fog computing in relaying networks

in Fig. 16. Compared to the traditional centralized storage/caching mechanism, distributed relay caching aims to achieve a trade-off between transmission rate and storage; relay caching thus can relax the traffic burden on the backhaul and backbone links and reduce the content delivery latency.

6.2 Video Streaming Application

Video streaming services can be either live or on-demand streaming. In live streaming, the user streams real-time content, such as a news channel; in on-demand streaming, the user orders specific content from a library content list. In this work, a video on-demand streaming service is considered. The scale of content acquired by service providers is growing significantly, although it is not necessary to stream all content from the cloud server, as this overloads the backhaul links and significantly increases the end-to-end delay. In distributed caching relay architecture, some popular content is therefore stored locally in relay stations to decrease the burden on the backhaul links, relax the traffic at the cloud servers, and provide fast content access.

6.3   Simulation and Results

The performance of the distributed relay caching architecture is evaluated with a scenario composed of one BS overlaid with up to eight RSs to form a multi-hop relay network, as shown in Figs. 17 and 18. Each BS/RS station serves a set of users who are uniformly distributed in the area with a random mobility model. Table 2 presents the parameters of the multi-hop relay network simulation.

The traffic model used was characterized by the parameters in Table 3. Figures 18, 19 and 20 show various performance results and emphasizes the effectiveness of our proposed distributed architecture versus the traditional centralized one.

Figure 18 shows the average achieved throughput among users for different architectures. It can be clearly seen from Fig. 18 how relays can be used to enhance the throughput and extend the coverage areas of BS. When compared to the conventional architecture without relay, the two-hop relay architecture provides the best performance in terms of throughput, followed by the single-hop relay architecture.

Figure 19 shows the packet loss for different architectures. The number of RSs does have a direct effect on the packet loss metric, where it can be seen that the packet loss rates of conventional and single-hop relay architectures tend to be similar

Fig. 17 System model

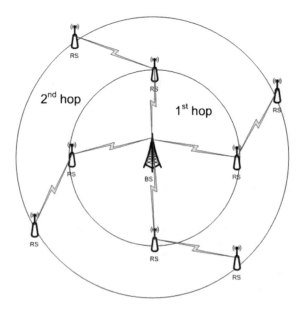

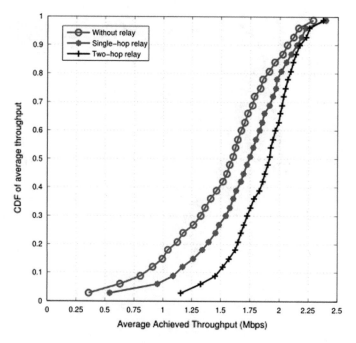

Fig. 18 Average achieved throughput for different architectures

Table 2 Simulation parameters

Parameter	Value
Base station transmit power	43 dBm
Relay station transmit power	30 dBm
Base station bandwidth	20 MHz
Relay station bandwidth	5 MHz
Size of relay caching	100 GB

Table 3 Video application parameters

Parameter	Value
Minimum video flow	1.5 Mbps
Size of the video content	1 GB
Number of video content	1000 title
Maximum delay tolerated	80 ms
Acceptable maximum packet lost	10%

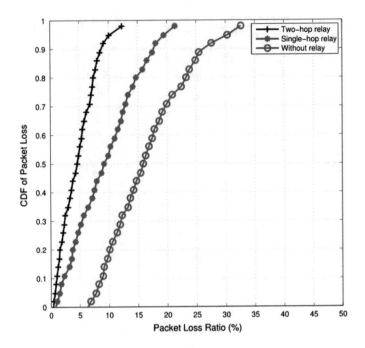

Fig. 19 Packet-loss for different architectures

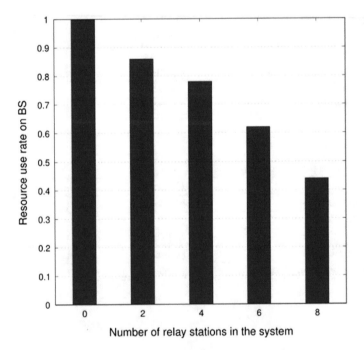

Fig. 20 Resource use rate on BS per RS concentration

(with high and medium packet loss, respectively) as compared to the two-hop relay architecture, which clearly minimizes the lost packet. This occurs since relay stations of the second hops are usually deployed at the boundaries of the base station coverage area to fulfill the throughput requirements of the users at the cell edges.

Figure 20 shows how the BSs resource use rate is affected by the number of RSs. It can be seen that the increment in RSs decreases the utilization of the direct and backhaul access links of the BS.

The overall results make it clear that the proposed architecture offers better performance than the traditional, centralized cloud architecture, because it takes into consideration the relay stations positions and user distribution in its vicinity, along with caching and computation capabilities. However, in order to exploit the distributed architectures benefits, a few key factors should be considered and jointly optimized, such as the intelligent caching resource allocation strategies and cooperative caching policies among relay stations. The caching policies that are used–deciding what to cache and when to release caches in different relay stations–are crucial for improving overall caching performance. The traditional caching policies, such as first-in first-out, least recently used, and least frequently used, should be developed further to appropriately improve the cache hit ratio in fog computing architecture.

7 Conclusion

In order to better support IoT applications, five key deployment features in LTE-A systems were discussed in this article. Each brings certain advantages to wireless mobile networks, for both operators and users. Multi-hop relay can provide great advantages by extending coverage to mountainous and sparsely populated areas, which other kinds of base stations cannot do. Heterogeneous relay can improve coverage while saving bandwidth on the access link. Mobile relay enhances the data rate of high-speed mobile users. Multiple backhauling can provide higher peak data rates and improvements for cell edge users experience. Relay-assisted D2D offloads the serving base station as well as enhancing the average data rate. These deployment features are likely to be enhanced further and considered in 5G networks and future wireless mobile technologies. The chapter also presented a few challenges in lower layers to achieving the benefits and discussed the required modifications. The chapter simulated a video streaming scenario, and presented numerical results in order to demonstrate the validity of the proposed architecture.

References

1. Bhat, P., Nagata, S., Campoy, L., Berberana, I., Derham, T., Liu, G., Shen, X., Zong, P., Yang, J.: LTE-advanced: an operator perspective. IEEE Commun. Mag. **50**(2), 104–114 (2012)
2. BenMimoune, A., Khasawneh, F.A., Kadoch, M., Sun, S., Rong, B.: Inter-cell handoff performance improvement in LTE-a multi-hop relay networks. In: Proceedings of the 12th ACM International Symposium on Mobility Management and Wireless Access, pp. 112–117 (2014)

3. Yang, Y., Honglin, H., Jing, X., Mao, G.: Relay technologies for WiMax and LTE-advanced mobile systems. IEEE Commun. Mag. **47**(10), 100–105 (2009)
4. Loa, K., Wu, C-C., Sheu, S.-T., Yuan, Y., Chion, M., Huo, D., Xu, L.: IMT-advanced relay standards (WiMAX/LTE Update). IEEE Commun. Mag. **48**(8), 40–48 (2010)
5. Zhuang, W., Ismail, M.: Cooperation in wireless communication networks. IEEE Wirel. Commun. **19**(2), 10–20 (2012)
6. Ismail, M., Zhuang, W.: A distributed multi-service resource allocation algorithm in heterogeneous wireless access medium. J. Sel. Areas Commun. **30**(2), 425–432 (2012)
7. Le, L., Hossain, E.: Multihop cellular networks: Potential gains, research challenges, and a resource allocation framework. IEEE Commun. Mag. **45**(9), 66–73 (2007)
8. Akyildiz, I.F., Gutierrez-Estevez, D.M., Reyes, E.C.: the evolution to 4G cellular systems: LTE-advanced. J. Phys. Commun. **3**(4), 217–244 (2010)
9. Lin, Y.-D., Hsu, Y.-C.: Multihop cellular: a new architecture for wireless communications. In: Proceedings of IEEE Nineteenth Annual Joint Conference on Computer and Communications Societies, vol. 3, pp. 1273–1282 (2000)
10. Iwamura, M., Takahashi, H., Nagata, S.: Relay technology in LTE-advanced. Technol. Rep., NTT Docomo Tech J. **12**(2), 29–35 (2010)
11. BenMimoune, A., Kadoch, M.: Multi-hop relays for LTE public safety network. In: Proceeding of 13th Conference on Applied Informatics and Communications, pp. 201–206 (2013)
12. Khasawneh, F.A., BenMimoune, A., Kadoch, M., Osama S.B.: Intra-domain handoff management scheme for wireless mesh network. In: Proceedings of 8th Conference on Circuits Systems, Signal and Telecommunications, pp. 101–105 (2014)
13. Hu, H.: 3GPP TR 36.836. Evolved Universal Terrestrial Radio Access, Study on Mobile Relay, pp. 212–220 (2012)
14. Sui, Y., Vihriala, J., Papadogiannis, A., Sternad, M., Yang, W., Svensson, T.: Moving cells: a promising solution to boost performance for vehicular users. IEEE Commun. Mag. **51**(6), 62–68 (2013)
15. Hasan, M., Hossain, E.: Resource allocation for network-integrated device-to-device communications using smart relays. In: Proceedings of IEEE Globecom Workshops, pp. 591–596 (2013)
16. Teyeb, O., Van Phan, V., Raaf, B., Redana, S.: Dynamic relaying in 3GPP LTE-advanced networks. EURASIP J. Wirel. Commun. Netw. **2009**(6), 124–134 (2009)
17. Benmimoune, A., Khasawneh, F.A., Kadoch, M.: Resource allocation framework in 5G multi-hop relay system. In: Proceedings of IEEE Global Communications Conference, pp. 2345–2351 (2015)
18. Cluster, R. A. S.: 5G radio network architecture. Tech. Rep. Radio Access SpectrumFP7 Future Int. Clust. **13**(2), 34–39 (2011)
19. Wei, H.Y., Rykowski, J., Dixit, S.: Wireless relay networking using IEEE 802.16 wimax technologies. In: Proceedings of WiFi, WiMAX, and LTE Multi-Hop Mesh Networks: Basic Communication Protocols and Application Areas, pp. 223–227 (2013)
20. Yi, S., Li, C., Li, Q.: A survey of fog computing: concepts, applications and issues. In: Proceedings of the Workshop on Mobile Big Data, pp. 867–873 (2015)

Two-Way Authentication for the Internet-of-Things

Corinna Schmitt, Thomas Kothmayr, Wen Hu and Burkhard Stiller

Abstract This chapter introduces the first fully implemented two-way authentication security scheme for Internet-of-Things (IoT) based on existing Internet standards, specifically the Datagram Transport Layer Security (DTLS) protocol. By relying on an established standard, existing implementations, engineering techniques, and security infrastructure can be reused, which enables an easy security uptake. The proposed security scheme uses two public key cryptography algorithms, RSA (Rivest, Shamir und Adleman) and Elliptic Curve Cryptography (ECC), tailored for the resource heterogeneous nature of IoT devices. The two-way authentication solution presented is designed to work over standard communication stacks that offer UDP/IPv6 networking for Low power Wireless Personal Area Networks (LoWPANs). A prototype implementation of DTLS is presented here in the context of a system architecture, and the scheme's feasibility (low overheads and high interoperability) is demonstrated through extensive evaluations on the DTLS-supporting platform OPAL as clusterhead with children of different IoT hardware platforms.

C. Schmitt (✉) · B. Stiller
Communication Systems Group CSG, Department of Informatics IfI, University of Zurich,
Binzmühlestrasse 14, 8050 Zurich, Switzerland
e-mail: schmitt@ifi.uzh.ch

B. Stiller
e-mail: stiller@ifi.uzh.ch

T. Kothmayr
Fakultät Für Informatik, Technische Universität München, Boltzmannstrasse 3,
85748 Garching, Germany
e-mail: kothmayr@in.tum.de

W. Hu
School of Computer Science and Engineering, The University of New South Wales, Sydney,
NSW 2052, Australia
e-mail: wen.hu@unsw.edu.au

© Springer International Publishing AG 2017
D.P. Acharjya and M. Kalaiselvi Geetha (eds.), *Internet of Things:
Novel Advances and Envisioned Applications*, Studies in Big Data 25,
DOI 10.1007/978-3-319-53472-5_2

1   Introduction

Today, a multitude of envisioned as well as implemented use cases for the Internet-of-Things (IoT) and Wireless Sensor Networks (WSNs) exists. It is desirable for certain scenarios to also make data globally accessible: (a) to authorized users only and (b) to data processing units through the Internet. Naturally, much of the data collected, such as locations and personal identifiers, are of sensitive nature. Even seemingly inconspicuous data, such as the energy consumption measured by a smart meter, can lead to potential infringements on the users' privacy, e.g., by allowing an eavesdropper to conclude whether or not a user is currently at home. From an industry perspective a pressing need for security solutions based on standards has risen. The market research firm Gartner states in [1]:

> *"The Internet of Things concept will take more than 10 years to reach the Plateau of Productivity - mainly due to security challenges, privacy policies, data and wireless standards, and the realization that the Internet of Things requires the build-out of a topology of services, applications and a connecting infrastructure."*

Regarding the infrastructure, security risks are aggravated by the trend towards a separation of sensor network infrastructure and applications [2, 3]. Therefore, a true end-to-end security solution is required to achieve an adequate level of security for the IoT. Protecting data once it leaves the local network is not sufficient.

A similar scenario in the traditional computing world comprises a user browsing the Internet on top of an unsecured Wireless LAN (Local Area Network). Attackers in physical proximity of the user can capture the traffic between the user and a Web server. Well known countermeasures against such attacks include the establishment of a secured connection to the Web server via HTTPS (Secured Hyper Text Transfer Protocol), the use of a VPN (Virtual Private Network) tunnel to securely connect to a trusted VPN endpoint, and using wireless network security such as WPA (Wireless Protected Access). These solutions are comparable to security approaches in the IoT area. Using WPA is similar to the traditional use of a link layer encryption. The VPN solution is equivalent to creating a secure connection between a sensor node and a security endpoint, which may or may not be the final destination of the sensor data. Establishing a HTTPS connection with the server is comparable to our approach: We investigate the use of the DTLS protocol in an end-to-end security architecture for the IoT. DTLS is an adaption of the widespread TLS protocol, used to secure HTTPS, for unreliable datagram transport.

However, the Internet is not limited to servers, routers, and computers with manifold resources anymore as more constrained devices are connected to it, forming the Internet-of-Things (IoT). Those devices - sensor nodes/motes - are very limited in memory (approx. 10–50 kByte RAM and 100–256 kByte ROM), computational capacity, and power (a few AAA batteries). Nevertheless, these devices still have to support end-to-end security, as requested by IoT, and secure communication with their limited resources.

For developing a security solution developers have to take design decisions into account. The presented solution in this book chapter considers three essential high-level design decisions:

Implementation of a standards-based design: Standardization has helped the widespread uptake of technologies. Radio chips can rely on IEEE 802.15.4 for the physical and the MAC layer. The IPv6 Routing Protocol for Low power and Lossy Networks (RPL), or called IPv6 over Low Power Wireless Personal Area Networks (6LoWPAN), provides routing functionality and the Constrained Application Protocol (CoAP) [4, 5] defines the application layer. So far, no such efforts have addressed security in a wider context for the IoT.

Focus on application layer end-to-end security: An end-to-end protocol provides security even if the underlying network infrastructure is only partially under the user's control. As the infrastructure for Machine-to-Machine (M2M) communication is getting increasingly commoditized, this scenario becomes more likely: The European Telecommunications Standards Institute (ETSI) is currently developing a standard that focuses on providing a "horizontal M2M service platform" [2], meaning that it plans to standardize the transport of local device data to a remote data center. For stationary installations security functionality could be provided by the gateway to the higher-level network. However, such gateways would present a high-value target for an attacker. If the devices are mobile, for example in a logistics application, there may not be a gateway to a provider's network that is under the user's control, similar to how users of smart phones connect directly to their carrier's network. Another example that favors end-to-end security is a multi-tenancy office building that is equipped with a common infrastructure for metering and climate-control purposes. The tenants share the infrastructure but are still able to keep their devices' data private from other members of the network. Using a protocol like Datagram Transport Layer Security (DTLS), which is placed between transport and application layer, does not require that the infrastructure provider supports the security mechanism. It is purely in the hands of the two communicating applications to establish security. If the security is provided by a network layer protocol, such as IPsec (Internet Protocol Security), the same is true to a lesser degree, because the network stacks of both devices must support the same security protocol.

Support for UDP (User Datagram Protocol): Reliable transport protocols like the Transmission Control Protocol (TCP) incur an overhead over simpler, unreliable protocols such as UDP. Especially for energy starved, battery powered devices this overhead is often extremely costly and TCP has been shown to perform poorly in low-bandwidth scenarios [6]. This is reflected in the design of the emerging standard CoAP, which uses UDP transport and defines a binding to DTLS for security [5]. By using DTLS in conjunction with UDP in the proposed approach (cf. Sect. 4) the application developer are not forced to use reliable transport as would be the case if TLS would be used. It is still possible to use DTLS over transport protocols like TCP, because DTLS only assumes an unreliable transport.

A resource-full device, perhaps including a Trusted Platform Module (TPM), raises the memory capacity to 50 kByte RAM and 256 kByte ROM, which allows the use of the DTLS protocol in an end-to-end security architecture for IoT. DTLS is an adaption of the widespread TLS (Transport Layer Security) protocol, used for Hypertext Transfer Protocol Secure (HTTPS), for unreliable datagram transport. In the proposed solution for resource-full devices a common DTLS handshake is performed, where both communication parties authenticate each other using a X.509 certificate in order to establish a secure communication channel before exchanging data itself. When looking on third design decision it is a weaker property than the reliability provided by TCP. However, the adaptations of DTLS for an unreliable transport introduces additional overhead compared to TLS. It might be beneficial to use TCP during the handshake phase, but the DTLS reliability mechanism for the proposed solution should be adapted to the special requirements of constrained networks as it is the case for wireless sensor networks. A study of TCP's influence handshake is, therefore, out of scope of this article. For further details it is referred to [7].

For devices with fewer resources (e.g., at around 10 kByte RAM and 100 kByte ROM), two-way authentication, as done within DTLS, is not feasible, because computing resources are too limited for extensive computations. But the aforementioned three high-level design decisions should also be supported. Therefore, Noack [8] developed a similar two-way authentication handshake solution applying keyed hash functions for authorization and message authentication as proposed by [9], instead of certificates. For further information it is referred to [8].

The rest of this book chapter is organized as follows: First, background information about the IoT, the specialized area Wireless Sensor Networks, and device class definition is presented. Followed by a brief characterization of existing security solutions for different devices classes. Section 3 introduces the standard-based end-to-end security architecture in detail, as well as general security goals and device's roles in a WSN. For resource-full devices a DTLS-based solution is presented in Sect. 4 based on [7]. The solution supports two-way authentication for all involved communication parties. Section 5 focuses on evaluation of the proposed solution addressing resource consumption, handshake performance, and comparison to related work. Finally, a brief summary and conclusion is given.

2 Background Information

This section presents the principles of the IoT with a special focus of WSNs and the used hardware. Furthermore, the vulnerabilities of such networks to different attacks are presents as well as the resulting necessary end-to-end security. Finally, related work for the proposed DTLS-based solution is presented.

2.1 Wireless Sensor Networks and the Internet-of-Thing

WSNs consist of a large number of small, cheap, and smart computing and sensing devices connected over radio. The advances in hardware, software, and networking have made practical WSN deployment technically and economically viable and enabled many applications in the areas such as habitat monitoring, health, structure monitoring, precision agriculture, and military [10].

While WSNs have revolutionized the way in which the authors of this book chapter understand, monitor, and control complex physical environments, the communication protocols for WSNs were custom-made to minimize resource consumptions for embedded sensing devices. Recent research in the IoT area has been in favor of the unique identification (e.g., by IPv6 addresses) of embedded computing devices within the existing Internet infrastructure over resource consumption. Therefore, the embedded devices in IoT are globally addressable and communicate with optimized Internet communication protocols such as 6LoWPAN. The IoT facilitates machine-to-machine communications (M2M) without human intervention and enables advanced applications such as smart home control and wearable computing. For further information about the IoT it is referred to [11, 12]. Within this book chapter the authors focus on WSNs, especially on light-weighted solutions, due to the limited hardware resources of used devices. Detailed information about requirements, protocols, and architecture requirements are presented in [10]. In the following only a brief overview of device resources are introduced in order to justify the proposed solution in Sect. 4.

As mentioned before the used devices are small; thus, the devices have limited resources that gave the devices the name: "constrained devices". They are limited in memory, computational capacity, and power. In the RFC 7228 [13] constrained devices were specified and organized in classes based on memory resources. These devices can be grouped into the following three classes:

- **Class 0 devices** are sensor-like nodes, usually pre-configured, and have less than 10 kByte RAM respectively 100 kByte Flash. In general class 0 devices are not able to communicate directly and secure with the Internet. In order to participate in Internet communications help of larger devices is required. Looking on the aforementioned appications' scope class 0 devices are unable to secure or managed communication in traditional sense.
- **Class 1 devices** have around 10 kByte RAM respectively 100 kByte Flash available. Compared to class 0 devices they are unable to talk easily to other Internet nodes employing a full protocol stack (e.g., HTTP, TLS, security protocols, Extensible Markup Language (XML) based data representations). Generally, class 1 devices use specifically designed protocol stacks (e.g., CoAP over UDP) and, therefore, do not require gateway nodes for conversation purposes. Class 1 devices are able to provide support for security functions required on large networks, as it is scope of this proposal. Furthermore, those devices can be integrated as fully developed peers into an IP network.

- **Class 2 devices** are more memory richer, usually having around 50 kByte RAM and 250 kByte Flash available. With this range they can support mostly same protocol stacks as used on notebooks or servers. But class 2 devices can still benefit from lightweight and energy-efficient protocols in order to increase lifetime. Compared to the other device classes those devices use a smaller percentage of their resources for networking, leaving more resources available for applications. Furthermore, if class 2 devices also support protocol stacks for lower classes development costs can be reduced and interoperability increased.

Devices with capabilities significantly beyond class 2 must be mentioned, because they are also included in the wide device diversity in the IoT. Those devices can usually use existing protocols in an unchanged manner, but may still be constrained by a limited energy supply. Smartphones are an example for these class 2+ devices.

Throughout this book chapter all devices within class 0 and class 1 build the group of "resource-less devices"; for all devices in class 2 or higher the term "resource-full device" is used.

As mentioned before, another limiting factor is energy. Reference [14] points out that power consumption of wireless sensor network devices could be divided into three domains: sensing, communication, and data processing. The latter two are also valid for devices beyond class 2. All three domains are closely related to the application and supported protocols in the network. If sensing takes place only sporadically, power consumption will be smaller than supporting constant monitoring tasks. Another dimension that requires energy is an event detection that is closely related to sleep modes. Most energy is required for communication purposes, including transmission and reception of data. Depending on the used power and data size/format the cost can differ. But the most consuming part - the start-up of the transceiver and, thus, turning a device on or off - does not only save energy. By looking on energy consumption for data processing it is smaller compared to communications and it is highly influenced by protocols supported. A good comparison is presented in [14]:

> *"Assuming Rayleigh fading and fourth power distance loss, the energy cost of transmitting 1 kByte a distance of 100 m is approximately the same as that for executing 3 million instructions by a 100 million instructions per second (MIPS)/W processor."*

They also pointed out that it is an advantage to integrated powerful devices in the network and outsource data processing functions to those devices if possible. As pointed out in [13] all devices can also be grouped corresponding to energy and power. In general, it can be said that devices beyond class 2 usually have unlimited power resource. Class 0, class 1, and class 2 devices can be distinguished in event energy-limited devices using event-based harvesting sources, period energy-limited devices including battery recharging or replacing, and lifetime energy-limited devices with fixed batteries without replacing them. The energy support is closely related to the application, its lifetime, and the type of power source used by the device.

2.2 Security Solutions for Wireless Sensor Networks

As described before, WSNs within the area of IoT are very prone security-wise, because the information transmitted includes sensitive data, is transmitted via the wireless medium using UDP, and usually only the end points of the communication chain should be able to read the message. Thus, the call for end-to-end security solutions grows. The development of a solution is challenging, because of the heterogeneous characteristics of deployed networks and the used devices with limited resources. Depending on the hardware used and the application itself different solutions were developed addressing resource-full and resource-less devices as pointed out in [7, 8].

Traditionally, security protocols in sensor networks focus on link layer security, protecting data on a hop-by-hop basis. The simplest approach to link layer security consists of using a network-wide encryption key, which often is the case in ZigBee networks [15]. ZigBee also provides support for cluster and individual link keys. MiniSec [16] is another well-known security mechanism for WSNs that provides data confidentiality, authentication and replay protection. Similar to ZigBee, the packet overhead introduced by MiniSec has only a few Byte. The widespread TinySec link layer security mechanism is no longer considered secure [16]. Most security protocols do not include a mechanism on how encryption keys are distributed to the nodes. Keys are either loaded onto the nodes before setup or a separate key establishment protocol is used. Public key cryptography (PKC) is used in traditional computing to facilitate secure key establishment. However, public key cryptography, in particular the widespread RSA algorithm, has been considered too resource consuming for constrained devices. Some security protocols, such as Sizzle [17], advocate the use of the more resource efficient ECC public key cryptosystem. Other research efforts, such as the secFleck [18] mote, provide support for faster RSA operations through hardware.

Approaches without PKC often rely on the pre-distribution of connection keys. Random key pre-distribution schemes, such as the q-composite scheme by [19], establish connections with a nodes neighbors with a certain probability $p < 1$. Intuitively, pre-distributed key schemes such as this require a large amount of keys to be loaded onto the nodes before deployment. Depending on the method used, this approach is scaling in $O(n^2)$ or $O(n)$ where n is the number of nodes in the network. The Peer Intermediaries for Key Establishment (PIKE) protocol achieves sub linear scaling in $O(\sqrt[3]{n})$ by relying on the other nodes as trusted intermediaries. While PIKE provides higher memory efficiency than random schemes, it still leaks additional key information when motes are captured.

Recently, more research into end-to-end security protocols for the IoT and WSNs is being conducted. As outlined in the introduction, such a protocol protects the message payload from the data source until it reaches its target. Because end-to-end protocols are usually implemented in the network or application layer, for-warding nodes do not need to perform any additional cryptographic operations since the routing information is transmitted in the clear. On the flip side, this means end-to-end security protocols do not provide the same level of protection of a network's availability as a link layer protocol. One example of an end-to-end security protocol is Sizzle by [17]. Sizzle is a compact web server stack providing HTTP services secured by SSL (Secure Sockets Layer). It uses 160-bit ECC keys for key establishment, which provide a similar level of security as 1024-bit RSA keys. In contrast to the presented solution in Sect. 4, it requires a reliable transport layer, which has been shown to incur large performance penalties in low bandwidth situations [6]. Sizzle also omits two-way authentication: Only the Sizzle enabled node is authenticated by a remote, more resource rich, client. This is insufficient for machine-to-machine communication in the IoT. SSNAIL [20] makes similar design choices as Sizzle and performs an ECC handshake over reliable TCP transport. Similar to the implementation of the described DTLS solution in this book chapter, SSNAIL is able to perform a full, two-way authenticated handshake but it still requires a reliable transport protocol.

Voigt Raza et al. [21] discussed how the IPsec protocol could be integrated into 6LoWPAN, the compressed IPv6 implementation used in most IP-enabled sensor networks. Their work focuses on how data transfer with IPsec can be made efficient in the context of 6LoWPAN. Regarding the Internet Key Exchange (IKE) protocol, which is used for key establishment in IPsec networks, methods for reducing the headers to make IKE more suitable for constrained devices is discussed [22], but authors do not present a performance analysis alongside their proposal.

As mentioned in Sect. 1, CoAP is an application layer standardization effort for the IoT. The current draft specifies a binding of CoAP to DTLS to achieve security [5]. Another proposal [23] aims to reduce the communication overhead of the DTLS headers through compression. As with the work on IPsec, the authors of this book chapter are currently not aware of any publication evaluating the performance of DTLS over 6LoWPAN. The presented DTLS solution in this book chapter can thus support these efforts by providing a set of real-world measurements from the presented DTLS implementation.

3   A Standard-Based End-to-End Security Architecture

The assumed system architecture is following the IoT model. It is assumed that IPv6 connects the Internet in the near future and parts of it run 6LoWPAN. The Transport layer in 6LoWPAN is UDP, which can be considered unreliable; the routing layer is RPL [24] or Hydro [6]. The DTLS implementation uses Hydro for routing, because at the time of writing the implementation code there was no available RPL implementation for TinyOS. RPL has since been standardized in RFC 6550 [24] and is distributed with newer versions of TinyOS. Thus, RPL was chosen for the two-way authentication solution with ECC for resource-less devices that is not addressed in

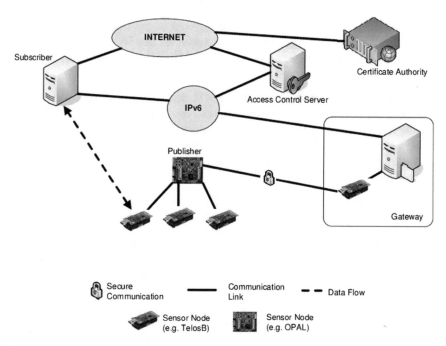

Fig. 1 System architecture

this book chapter but can be read in detail in [8]. However, both routing protocols are similar enough so that a change should have negligible impact on the presented results. IEEE 802.15.4 is used for the physical and Media Access Control (MAC) layer. Based on this protocol stack DTLS and ECC were chosen as security protocol, which places it in the application layer on top of the UDP transport layer. The final stack structure might differ depending on chosen deployment, application, and implementation of functionalities. For example, the certificate authority (CA) or Access Control (AC) server can be included in the gateway complex. Detailed description is available in [7, 25].

The general idea of the architecture (cf. Fig. 1 [7]) is that a subscriber wants to access data of the WSN. A designated device in the WSN, called publisher, is allowed to publish collected data and make it available from outside of the WSN (e.g., for analysis purposes or announcement to other systems like an air conditioning system). The subscriber has to establish a connection to the publisher and, therefore, it is necessary to establish an end-to-end secured connection. Because of sensitive data (e.g., Global Positioning System (GPS) data) included in the measurements of sensor nodes it is essential to authenticate the communication partners (here: subscriber and publisher). Depending on the resources of the devices different solutions are possible. This book chapter focuses on resource-full devices and the solution is described in upcoming Sect. 4 For an appropriate solution for resource-less device it is referred to [8, 26].

Due to the assumption that WSNs are a subset of the IoT area, it is assumed that similar security needs as in traditional networks, like IP networks, have to be considered. Thus, the solution proposed will address the following three security goals:

- **Authenticity**: Recipients of a message can identify their communication partners and can detect if the sender information has been forged.
- **Integrity**: Communication partners can detect changes to a message during transmission.
- **Confidentiality**: Attackers cannot gain knowledge about the contents of a secured message.

For the remainder of this book chapter the following roles for devices in a WSN are considered, all besides the roles of publisher and subscriber as introduced in [8]:

- **Collectors** are limited to collecting and transmitting environmental data. They do not execute any preprocessing tasks on collected data; instead they send only raw information. Those could include, for example, humidity, temperature, light, and voltage. Measurements are executed periodically and immediately followed by data transmission to the gateway.
- **Aggregators** are selected devices in the network with more resources than collectors. The aggregator can pre-process data within the network and, thus, reduce the traffic in the network. The performed degree of aggregation (doa) can vary depending on the device resources and per-formed function. Examples are data aggregation and message aggregation as described in [25].
- The **gateway** is usually a complex of a sensor node and a server-like component (e.g., router, server, persona computer (PC)). It connects the WSN components to the IP networks on the outside. It basically brings wireless communication to wired communication and makes the collected data available for other applications [25].

4 DTLS Solution for Resource-Full Devices

This section assumes resource-full devices that are able to perform difficult encryption operations and, thus, can perform a DTLS handshake in order to support two-way authentication. In the following the used message structure, the performed DTLS handshake, and security considerations are addressed following [7, 25, 26].

4.1 DTLS Protected Message Structure

All messages sent via DTLS are prepended with a 13 Byte long DTLS record header. This header specifies the content of the message (e.g., application data or handshake

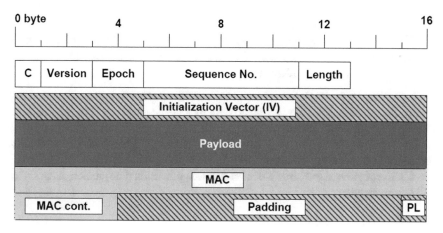

Fig. 2 A DTLS record protected with CBC block cipher

data), the version of the protocol employed, as well as a 64-bit sequence number and the record length. The top two Byte of the sequence number are used to specify the epoch of the message, which changes once new encryption parameters have been negotiated between client and server. Figure 2 shows the DTLS record header in white [7]. The record header is either followed by the plaintext if no security has been negotiated yet, or by the DTLS block cipher. If a block cipher is used, the plaintext is prepended by a random Initialization Vector (IV), which has the size of the cipher block length. This protects against attacks where attackers can adaptively choose plaintext. The plaintext is followed by a Hash-based Message Authentication Code (HMAC), which allows the receiver to detect if the DTLS record has been altered. Finally, the message is padded to a multiple of the cipher block length. The payload itself is encrypted with the block cipher, where IV and padding are not used to calculate the HMAC. Unlike TLS, DTLS does not allow for stream ciphers because they are sensitive to message loss and reordering. Instead, DTLS uses block ciphers in the Cipher-Block Chaining (CBC) mode of operation.

4.2 Certificate Structure

As mentioned earlier the proposed solution in this book chapter requires certificates for authentication purposes. Thus, this section briefly introduces the general structure of X.509 certificates based on RFC 6818 [27]. For further information it is referred to the RFC 6818 and common literature about public key infrastrukture (PKI). Based on RFC 6818 a X.509 certificate should include the following items in general:

1. Serial Number
2. Validity: Not Before: Date and time, Not After: Date and time

3. Subject: commonName = localhost
4. X509v3 extensions including X509v3 Basic Constraints: CA:FALSE, Netscape
   Comment: OpenSSL Generated Certificate, X509v3 Subject Key Identifier, and
   X509v3 Authority Key Identifier

Depending on the implementation additional information should be requested that will be incorporated into the certificate request. For the proposed solution the items that where selected are country name (2 letter code), state or providence name (full name), locality name (e.g., city), organization name (e.g., company), organization unit name (e.g., section), common name (e.g., name), Email address, challenge password, and company name. Additional and optional items can be challenge password and company name.

4.3 DTLS Handshake Assumption

The key material and cipher suite, consisting of a block cipher and a hash algorithm, are negotiated between client and server during the handshake phase, which commences before any application data can be transferred. There are three types of handshake: unauthenticated, server authenticated and fully authenticated handshakes. During an unauthenticated handshake neither party authenticates against the other and during a server-authenticated handshake only the server proves its identity to the client. In a fully authenticated handshake the client has to authenticate itself to the server as well. In the following the unauthenticated handshake is not considered, because it provides no authenticity at all.

There are different algorithms that can be used for authentication in a DTLS handshake. Variants based on ECC have been shown in embedded networks [17]. Since it was argued for standard-based communication architecture for the IoT to promote interoperability, the rest of this section will focus on authentication based on RSA. Because it is todays dominant PKC system [28] a suitable infrastructure for obtaining certificates from commercial certificate authorities is already in place.

Figure 3 shows a fully authenticated DTLS handshake [7]. Individual messages are grouped into "message flights" according to their direction and occurrence sequence. Flight 1 and 2 are an optional feature to protect the server against Denial-of-Service (DoS) attacks. The client has to prove that it can receive as well as send data by resending its *ClientHello* message with the cookie sent in the *ClientHello-Verify* message by the server. The *ClientHello* message contains the protocol version supported by the client as well as the cipher suites that it supports. The server answers with its *ServerHello* message that contains the cipher suite chosen from the list offered by the client. The server also sends a X.509 certificate to authenticate itself followed by a *CertificateRequest* message if the server expects the client to authenticate. The *ServerHelloDone* message only indicates the end of flight 4. If requested and supported, the client sends its own certificate message at the beginning of flight 5. The *ClientKeyExchange* message contains half of the pre-master secret encrypted

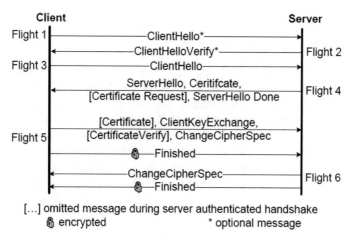

Fig. 3 Fully authenticated DTLS handshake

with the servers public RSA key from the server's certificate. The other half of the pre-master secret was transmitted unprotected in the *ServerHello* message. The keying material is subsequently derived from the pre-master secret. Since half of the pre-master secret is encrypted with the servers public key it can only complete the handshake if it possesses the private key matching the public key in the server certificate. Accordingly, in the *CertificateVerify* message the client authenticates itself by proving that it is in possession of the private key matching the client's public key.

It does this by signing a hashed digest of all previous handshake messages with its private key. The server can verify it through the public key of the client. The *ChangeCipherSpec* message indicates that all following messages by the client will be encrypted with the negotiated cipher suite and keying material. The *Finished* message contains an encrypted message digest of all previous handshake messages to ensure both parties are indeed operating based on the same, unaltered, handshake data. The server answers with its own *ChangeCiperSpec* and *Finished* message to complete the handshake.

4.4 Security Considerations for DTLS Handshake

By choosing DTLS as the security protocol the aforementioned security goals can be achieved. DTLS is a modification of TLS for the unreliable UDP and inherits its security properties [29]. Using an application layer security protocol like DTLS as opposed to link or network layer security protocols such as MiniSec [16] has a number of advantages but also some drawbacks as described in the following.

Lower layer security protocols do not provide end-to-end communication security. Data is decrypted on receipt and re-encrypted for forwarding on each hop in a multi-hop network. An attacker can therefore gain access to all clear text data that passes through a compromised node. Scalability is often also an issue for these protocols because they need to establish a secured connection with each of their neighbors to form a mesh network and cryptographic overhead occurs on each hop. On the other hand, in an end-to-end security protocol, cryptographic overhead occurs on the sender and receiver only. Compromised nodes only provide an attacker with access to the measurement data from local nodes. Routing algorithms are also agnostic of the payload protection, thus even nodes that have not established a secure connection can be used to forward packets to a subscriber/destination. A scenario could be in an office building shared by multiple occupants (parties): each party subscribes to a part of the sensor readings only and wishes to keep the data they subscribed to private from other parties, yet they still may share a common communication network in order to reduce cost.

However, an application layer security protocol does not protect routing information. Adversaries can therefore analyze the traffic patterns of a network in clear text. They may even launch a DoS, worm hole, or resource consumption attack that lowers the availability of the network [30]. In this book chapter, the authors focus on end-to-end communication security and rely on other schemes for securing lower communication layers [30].

Scenarios like the one above raise the need for proper authentication of data publishing devices and access control throughout the network. Therefore, an AC was included into the assumed system architecture. The AC is a trusted entity and a more resource-rich server, on which the access rights for the publishers (= motes) of the secured network are stored. The identity of a default subscriber is usually preconfigured on a publisher before it is deployed. If any additional subscribers want to initialize a connection with the publisher, they first have to obtain an access ticket from the AC. The AC verifies that the subscriber has the right to access the information available from the publisher. The publisher then only has to evaluate the identity of the subscriber and verify the ticket it has received from the AC. Details of this scenario are subsequently omitted because they are out of scope of this chapter. More details can be found in [25]. This requires a unique identity for a publisher in the network. In the Internet, identities are usually established via PKC and the identifiers provided through X.509 certificates. A X.509 certificate contains, among other information, the public key of an entity and its common name (e.g., my-bank.com). An example is presented in [31]. A trusted third party - the CA -, which serves two purposes, signs the certificate: Firstly, the signature allows the receiver to detect modifications to the certificate. Secondly, it also states that the CA has verified the identity of the entity that requested the certificate.

Hu et al. showed that RSA, the most commonly used public key algorithm in the Internet, can be used in sensor networks with the assistance of a Trusted Platform Module (TPM), which costs less than 5% of a common sensor node [18]. A TPM is an embedded chip that provides tamper proof generation and storage of RSA keys as well as hardware support for the RSA algorithm. The certificate of a TPM equipped publisher and the certificate of a trusted CA must be stored on the publisher prior to deployment. For publishers that are not equipped with TPM chips an authentication via the DTLS pre-shared key cipher-suite is proposed, which requires a small number of random Byte, from which the actual key is derived, to be preloaded to the publishers before deployment. This secret must also be made available to the AC server, which will disclose the key to devices with sufficient authorization.

For the sake of completeness, it should be mentioned here that an alternative was also implemented using ECC to achieve two-way authentication for resource-less devices. ECC is used for key generation, key exchange, signatures, and encryption. The applied two-way authentication protocol is based on a modified version of the Bellare, Canetti, Krawczyk (BCK) handshake [32], where the modification consists of the usage of pre-shared keys for defense against a man-in-the-middle attack and for additional authorization of different communication parties. For more details it is referred to [8, 26].

5   Evaluation

The proposed DTLS solution was evaluated on test-beds including devices of class 1 (e.g., TelosB [33]) and class 2 (e.g., OPAL [34]). The setup was already described in Sect. 3 and the protocol was implemented in TinyOS 2.x with Berkeley Low-power IP stack (BLIP) version 2.0. As pointed out in Sect. 2, the main challenge for two-way authentication solutions is a resource efficient solution that requires only a small part of the available resources, allowing running a reasonable application in addition to the security solution. Thus, the following sections will focus on resource consumption and handshake performance, as well as on a comparison to the related work mentioned in Sect. 2.2 analog to references [7, 25, 35].

Previous work has already demonstrated techniques to reduce the protocol header overhead during data transmission [21] and has proven the feasibility of performing software encryption and hashing on the sensor node [16], also called mote. Indeed, Raza et al. recently have made first proposals for a compressed header format [23]. Gupta et al. showed the feasibility of a server authenticated SSL handshake [17]. Therefore, the component of the security architecture that is currently least under-stood in the context of the IoT is the fully authenticated DTLS handshake, which includes both client and server authentication.

We have implemented a DTLS client that performs the DTLS handshake with an OpenSSL 1.0.0d server. The client is targeted at the OPAL sensor node [34], which features an Atmel SAM3U micro-controller [36] and the Atmel AT97SC3203S TPM [37]. It has 48 kByte RAM and the micro-controller is clocked at 48 MHz in the implementation. In the following sections the implementation will be evaluated with regards to its performance during the handshake and data transmission, as well as its energy and memory consumption. Unless otherwise stated, the DTLS cipher suite performed was TLS-RSA-with-AES-128-CBC-SHA. AES-128 has been shown to be one of the fastest block ciphers on motes [38] and offers sufficient security. Furthermore, the selected cipher suite is the required block cipher suite for DTLS from version 1.2 onwards. Other common cipher suites are either based on RC4, which is a stream cipher and thus not permitted by DTLS, or 3DES, which is very slow and as a result causes a large cryptographic overhead.

5.1 Data Transfer Latency

In this section the latency as a measure of the systems cryptographic performance is considered. The round-trip time (RTT) for different sizes of plaintext data through a single hop network and a multi hop network with four hops was evaluated as shown in [7]. The timing for the DTLS packets on the mote was measured. Readings for pure plaintext data without any additional headers were obtained by issuing the ping6 command on the subscriber.

A packet sent with both a SHA-1 HMAC and AES-128 encryption is denoted as 'AES-128'. The denotation 'SHA-1' is used if a packet only contained a SHA-1 HMAC. The reading for 8 Byte plaintext data is missing, because the ICMP-Header and the timestamp sent by ping6 are together at least 16 Byte long.

The measurements show a linear increase of round-trip time with jumps occurring approximately every 100 Byte. These spikes can be attributed to the 128 Byte maximum link layer frame size defined by IEEE 802.15.4, which includes header and trailer. These jumps occur earlier when sending DTLS protected packets due to the additional DTLS packet headers, the HMAC size and the explicit Initialization Vector in each packet. See Sect. 4.1 for more details on the packet structure.

Both the increased packet size and processing overhead lead to increased end-to-end transmission latency for DTLS packets compared to plaintext packets (cf. Fig. 4 [7]). In the single hop scenario, transmission latency was increased by up to 95 ms for AES-128 and up to 75 ms for SHA-1 encryption, which were an average increase of 62% and 35% respectively over the plaintext case. In the multi hop

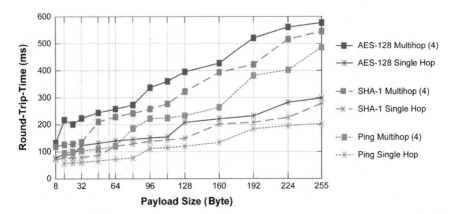

Fig. 4 Average (n = 100) packet round-trip time for different plaintext sizes

scenario, round trip times increased by a maximum of 163 ms and were 74% longer on average for AES-128 encrypted packets. Packets with a SHA-1 HMAC took up to 129 ms longer for the round-trip with an average of 40% more time being spent. The decreased performance for transmission latency is mostly due to the large packet overhead of up to 64 Byte, which consists of 13 Byte DTLS record header, 16 Byte Initialization Vector, 20 Byte HMAC, and up to 15 Byte padding. Calculating a SHA-1 hash of a 255 Byte plaintext message only takes 9 ms, encryption with AES-128 takes another 12 ms. Both operations do not contribute significantly to the overall transmission latency. This is consistent with the measurements for 16 Byte plaintext (RTT of 58 ms), which increases to 90 ms with AES-128. Including the overhead of the DTLS record format, 16 Byte plaintext are expanded to a 77 Byte message. Sending 80 Byte via ping requires 78 ms, which indicates a computational overhead of around 12 ms in this case. A more detailed analysis of the transmission overhead from an energy perspective is provided in Sect. 5.4.

5.2   Handshake Latency

Another performance indicator to consider is the latency introduced by performing a DTLS handshake. The time from the beginning of the establishment of the handshake was measured until a *Finished* message has been received on the client. In addition to using a 2048-bit key, the results for a 1024-bit key for comparison was included.

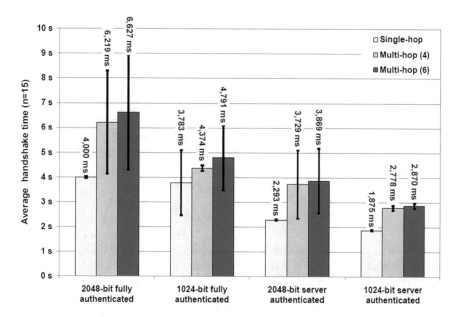

Fig. 5 Time to complete different types of DTLS handshakes

Figure 5 shows the average latency for a fully authenticated and a server-authenticated handshake [7]. For each type of handshake 15 measurements were conducted. The bars show the average over these measurements, and the error bars show the standard deviation.

The large standard deviation is caused by the presented implementation behavior when message loss occurs. DTLS states that an implementation should wait for an answer for a set amount of time after sending a flight of messages. If it does not receive an answer during this period, it retransmits the whole flight. This timeout value was set to 5 s to avoid unnecessary retransmissions in networks with a high end-to-end delay, which is common in a low power lossy network, and/or with energy limited thin clients that are slow to respond. DTLS implementations for the Internet often choose a retransmission timeout of 1 s or less. In general, it can be seen that the time to execute a handshake is shorter for smaller RSA keys and reduced by almost 2 s when client authentication is omitted in the handshake. It was observed that packet loss occurred mainly in a multi-hop environment and when larger DTLS messages were being sent. This increases the total handshake time significantly because of the large DTLS retransmission timeout. However, total energy consumption of the client does not increase significantly, because all TPM operations, which are the largest contributor to overall handshake energy costs, are only executed after successful receipt of all relevant server messages. Losing a packet with information obtained from the TPM does not lead to a repeated execution of the TPM operations because the resulting messages are buffered and can be retransmitted. During the experiments no failed handshake attempts where recognized.

DTLS requires successful transmission of all handshake packets over an unreliable transport layer. Since it provides its own reliability mechanism during the handshake, network topology, congestion and link quality have a large impact on the time needed to complete a DTLS handshake. One parameter the programmer can influence to achieve better performance in lossy networks is the maximum transmission unit (MTU) for DTLS handshake packets, which determines the size of individual handshake packet fragments. To study the influence of the MTU on overall handshake establishment time a random, artificial packet drop rate was introduced on the link layer and measured handshake completion times for various MTUs.

Figure 6 shows that even a small amount of packet loss has a large impact on overall handshake completion time [7]. It was considered that each link layer packet has an independent chance of being dropped, resulting in the total loss of all packets that follow. The probability for a packet loss is defined by

$$P \text{ (Packet loss)} = 1 - PLR^{\lceil \frac{traffic \ in \ Byte}{100 \ Byte} \rceil},$$

where PLR is the packet loss rate. If a typical, fully authenticated DTLS handshake was taken, which causes 2,438 Byte of traffic as an example, there is a 72.26% chance of packet loss while transmitting the 2,438 Byte of handshake payload at 5% link layer packet loss. If the link layer packet loss rate is 10%, there is a 92.82% chance of packet loss occurring. In that case, the DTLS reliability mechanism is waiting for a

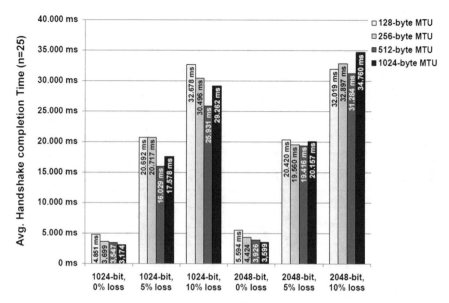

Fig. 6 Handshake completion times with various amounts of artificial link layer packet loss and different MTUs

timeout before resending the whole message flight [29]. As before, the retransmission timer was set to 5 s during the experiments.

The authors of this book chapter are considering uncorrelated packet loss in this evaluation; even tough packet loss is correlated in reality. The reasoning behind these figures is that it was unknown at which time during the handshake the interference that causes packet loss will start. Therefore, a constant probability of packet loss was used, which will cause all following fragments of the current message flight to be dropped. Additional, correlated packet loss before the next retransmission interval has no adverse impact, because the damage is already done.

The MTU influences the granularity at which handshake the receiver can reassemble messages. A small MTU splits large handshake messages into many different packets, allowing the receiver a fine-grained reassembly if packets are lost. Since every new packet has to bear the DTLS header, the overall amount of traffic increases, which in turn increases the probability of packet loss. A larger MTU splits messages into fewer packets, which reduces the probability of packet loss because there is less network traffic. However, if packet loss does occur, reassembly cannot be done as fine-grained as with a smaller MTU. Figure 4 shows that a MTU of 512 Byte seems to strike the best balance between reassembly and network traffic in the experiments.

Table 1 RAM and ROM usage by component

	RAM (Byte)	ROM (Byte)
Cryptography	541	10,838
DTLS messages	1,174	2,568
DTLS network	4,294	5,672
TPM	4,321	4,928
BLIP	6,352	9,298
Application	166	–
System	991	30,075
Total data + BSS	**17,839**	**63,379**
Stack minimum	1,098	0
Stack maximum	2,300	3,936
Total	**18,937 – 20,139**	**63,379 – 67,315**

5.3 Memory Consumption

In order to determine the static memory allocation of individual components of the presented implementation, the entries in the symbols table of the OPAL binary after compilation were analyzed. Memory has been measured for a fully authenticated handshake with 2048-bit RSA keys. This type of handshake has the largest memory requirements, because it needs more code and buffer space for the client's *Certificate* and *CertificateVerify* messages. The memory consumption was divided into six, respectively seven categories as illustrated in Table 1 [7]. Additionally, the maximum stack size was measured by filling the stack with a dummy variable directly after booting and analyzing how much of that continuous memory block had been overwritten after a successful DTLS handshake. Only the first subtotal of Table 1 [7], considers static memory allocation. Because it currently contributes a significant portion of overall stack use, two prototypical methods of initializing the client certificate were implemented. The method represented by Stack Minimum directly sets each individual Byte of the outgoing message buffer to the matching value from the Certificate. The drawback is a increased ROM use, because the code basically contains hundreds of statements in the form $buffer[x] = 0xff$. The "Stack Maximum" method initializes the outgoing message buffer from a temporary array, which is filled from a hardcoded, anonymous array, e.g., $uint8_t[CERTLEN] = \{0xff, 0xff, 0xff, \ldots\}$. In production the certificate would usually be read from the mote's flash memory, which should fall somewhere in between the figures from these two approaches.

In total, approximately 20 kByte of RAM and 67 kByte of ROM is required for the implementation. The BLIP implementation requires most of the resources, followed by TPM drivers and DTLS networking code. Overall, the implementation is still below the 48 kByte of RAM and 256 kByte of program memory provided by OPAL [7, 25].

5.4   Energy Consumption

The authors measured the energy consumption during the handshake phase across a $10\,\Omega$ resistor with an oscilloscope. It yielded a value for the electric potential, which can be converted into a value for the current draw by dividing it through the value of the resistance ($10\,\Omega$).

The energy costs can then be calculated as $U_{probe}/R * t * U_{battery}$. U_{probe} is the measured voltage, $R = 10\,\Omega$ is the value of the resistor, t is the transaction time, and $U_{battery} = 3.998$ V is the battery voltage. Table 2 shows the energy consumption during a typical execution of different handshake types [7]. A 2048-bit RSA key was used, because 1024-bit keys are not recommended for future deployments [39].

The contribution of the radio and micro-controller are neglected in further discussion. Both can be considerably reduced by using power saving techniques, e.g., by using the TinyOS Low Power Listening (LPL) Media Access Control layer for the radio (less than 1% radio duty cycles have been reported by the literature repeatedly) and setting the micro-controller into a lower power state where it consumes less than $15\,\mu A$ for SAM3U [36]. However, the transmission costs of messages increase significantly if LPL is activated. This tradeoff is subject to the design and configuration of each deployed network. For better comparison the idle energy use as outside of the developers' field of control was viewed and the focus was set on energy costs, which occurred in any case. Sending messages and performing cryptographic operations contribute very little (17.4 mJ and 4.18 mJ, respectively) to the overall energy costs that are directly dependent on the presented DTLS implementation. The total cost is then largely bound by the use of energy of the TPM.

Figure 7 shows the measured draw of the TPM chip [7]. "TPM Start" and "TPM Sign" are the longest consecutive operations, which consume 174.46 mJ and

Table 2 Transaction time/energy consumption of DTLS handshake (2048-bit key)

	Current (mA)	Fully authenticated handshake	Server-authenticated handshake
Computation	30	35 ms, 4.18 mJ	33 ms, 3.95 mJ
Radio TX	18	242 ms, 17.4 mJ	70 ms, 5.03 mJ
TPM Start	52.2	836 ms, 174.46 mJ	836 ms, 174.5 mJ
TPM TWI	43.6	688 ms, 120.0 mJ	476 ms, 83.0 mJ
TPM Verify	51.8	59 ms, 12.2 mJ	56 ms, 11.6 mJ
TPM Encrypt	51.8	39 ms, 8.07 mJ	40 ms, 8.28 mJ
TPM Sign	52.2	726 ms, 151.5 mJ	–
Total minimum		487.8 mJ	286.4 mJ
CPU idle	11.4	3965 ms, 180.7 mJ	2265 ms, 103.2 mJ
Radio idle	18	3758 ms, 270.4 mJ	2228 ms, 160.3 mJ
Total		939.0 mJ	549.9 mJ

Fig. 7 Current draw for a fully authenticated DTLS handshake

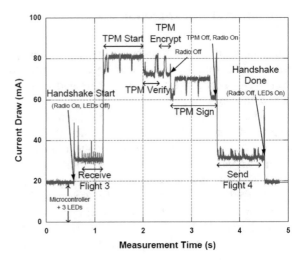

151.5 mJ. The TPM is performing an operation with its RSA private key in "TPM Sign", which is more complex than using a RSA public-key. During the "TPM Start" phase the TPM performs a series of internal self-tests to detect tampering and unauthorized commands. The second large block is "TPM TWI" which describes the amount of time that is spent passing data to the TPM and receiving data from it via the TWI bus clocked at 100 kHz, consuming 120 mJ. It shows as a lower current draw. It can be recognized that directly after the end of the "TPM Start" sequence and before the short spike in "TPM Verify". The spike is the actual verification operation performed by the TPM, which consumes only 12.2 mJ. Similarly, the actual "TPM Encrypt" operation is the spike that follows another section of data transfer on the TWI bus, consuming 8.07 mJ. During "TPM Verify" the TPM uses the stored key of a CA to verify the server certificate presented during the handshake. The "TPM Encrypt" operation is used to encrypt a nonce with the server's public key. If the mote is expected to authenticate itself during the handshake, it performs a "TPM Sign" operation to sign a hash over all previous handshake messages with its RSA private key. Since a server-authenticated handshake does not require the expensive "TPM Sign" operation, it uses significantly less energy but also provides weaker overall authentication because an attacker could impersonate a mote toward the server. Communication time is also shorter since the sensor node does not send its certificate [7, 25].

If two AA 280-mAh batteries power the mote, they contain approximately 30,240 Joule of energy. If 5% of the energy is used for DTLS handshakes for (re)keying purposes, which happen once per day, it could last for more than 8.5 years for a fully authenticated handshake at 487.8 mJ each, or more than 14.5 years for a server-authenticated handshake at 286.4 mJ each. As stated earlier, the calculation of a SHA-1 hash for 255 Byte takes 9 ms and encryption with AES-128 another 12 ms.

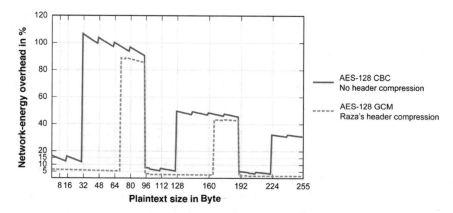

Fig. 8 Network energy overhead caused by the DTLS record format

Given the current draw for computation of 30 mA at 48 MHz clock speed from Table 2, this results in the order of 9.9 μJ per Byte [7, 25].

The energy consumption after the completion of the handshake is closely related to the latency values from Fig. 5, which portraits the influence of the network and processing overhead introduced by DTLS [7]. The increase in latency naturally also leads to an increase in energy consumption, because the radio has to be held in the transmitting state for a longer time in order to prevent it from entering a sleep state. Figure 8 shows the overhead in percent that occurs when a plaintext of a given size is encrypted and sent in a secure DTLS record [7]. The baseline for this comparison is the time it would take to send the plaintext without any additional headers or other meta data.

It is assumed that the energy cost to send a message with length x via BLIP follows a discontinuous piecewise linear function: $c(x, a, b) = \lceil \frac{x}{100} \rceil * a + x * b$. Here, a represents the amount of energy needed to access the medium for one IEEE 802.15.4 message and sending the preamble and all other fixed energy costs for one message. The energy required for transmitting one Byte of payload without fixed costs is represented by b. The constant 100 is the maximum link layer message length defined by BLIP. Since the relative overhead was only interesting for the authors, the current draw was ignored and only the relation between message length and time was analyzed. For this purpose the round-trip times were used as measured in Fig. 5 for a simple ping and divided them by two [7]. Matlab was used to find the minimum of the error function

$$err(a, b) = \sum_{x \in M} \| \frac{c(x, a, b) - t(x)}{x} \| \tag{1}$$

where M is the set of plaintext lengths for which we have obtained measurement times and $t(x)$ returns the measured time for a plaintext length x. This optimization returned $a = 27.368$ and $b = 0.072$. With these results the approximate time could be calculated that was required to send plaintext and larger DTLS records for the same amount of plaintext.

Figure 8 shows that the overhead introduced by the DTLS record format is under 17% for small plaintext lengths. It rises to over 100% when the DTLS record will not fit into a single link-layer packet anymore [7]. BLIP then has to fragment the packet and bear the expensive medium access a second time. One way to reduce the network overhead is reducing the size of DTLS records. The proposal is to employ the header compression detailed by [23]. It reduces the size of a DTLS record header from 13 to 5 Byte. Further savings are possible if the block cipher mode of operation is changed from CBC to Galouis/Counter mode of operation (GCM). The plaintext encrypted by GCM will always lead to a cipher text of the same length [40]. Since GCM belongs to the class of block cipher modes called Authenticated Encryption with Associated Data (AEAD) the SHA-1 HMAC is no longer necessary. Instead, GCM can be used directly to authenticate the data and associated headers. The maximum length GCM auth tag, which requires 16 Byte, thus replaces the 20-Byte SHA-1 HMAC. Additionally, the explicit IV is no longer necessary, because GCM is not susceptible to the vulnerability that makes the IV necessary. The maximum DTLS record overhead can thus be reduced from 64 Byte down to 21 Byte: Five Byte for the compressed record header plus the 16 Byte GCM auth tag. Figure 8 shows that this more than doubles the area in which a DTLS record only incurs little overhead over sending the plaintext directly [7].

In order to put the TPM energy consumption and processing time in context, measurements of RSA and ECC were also performed. The RSA and ECC TinyOS

Table 3 Software RSA (2048-bit key) on OPAL. One Private Key and two Public Key operations are required for a handshake

	Current (mA)	Computation time (ms)	Energy consumption (mJ)
RSA - Public Key @ 48 MHz	30	440	52.8
RSA - Private Key (high memory) @ 48 MHz	30	4,725	566.7
RSA - Private Key (low memory) @ 48 MHz	30	14,895	1,786
Handshake RSA total @ 48 MHz	**30**	**5,165**	**619.5**
RSA - Public Key @ 96 MHz	48	221	42.4
RSA - Private Key (high memory) @ 96 MHz	48	2,362	453.3
RSA - Private Key (low memory) @ 96 MHz	48	7,447	1,429
Handshake RSA total @ 96 MHz	**48**	**2,583**	**495.7**

modules available did not support 2048-bit RSA keys or their respective ECC equivalent. Thus, the authors ported the RSA and ECC implementation of the open source project CyaSSL [41] to TinyOS. This port includes many of the optimization techniques adopted in TinyECC [42], such as Barrett Reduction, Sliding Window multiplication, Shamir's Trick and others. It does not, however, include inline assembly instructions to speed up natural number operations. The presented implementation [43] is made available to the TinyOS community under the GPLv2 license [44]. Table 3 shows the results for individual RSA operations with a 2048-bit RSA key performed in software [7]. The figures for the handshake only pertain to the DTLS client, as was the case in the previous evaluations.

With a clock speed of 48 MHz, the software implementation requires more than twice as much time as the TPM and almost 1.5 times the amount of energy. The respective values for the TPM where 2348 ms and 466.2 mJ. This advantage is demised when the TPM is compared to software RSA being performed at 96 MHz, where both require roughly the same amount of time and energy. The RSA implementation still has room for improvement through embedded Assembler code and could thus be made more time and energy efficient than the TPM on the chosen platform. However, the TPM still provides secure storage of the RSA key, which cannot be achieved by software means, and the implementation complexity and RAM requirements of the TPM drivers are far less than those of software RSA implementation. Additionally, newer versions of the available TPM chip have more than halved the computation time for 2048-bit RSA keys.

If secure storage of a motes private key is not a design goal, a software implementation of ECC is recommended instead. As Table 4 shows, it requires far less time and energy than either solution for RSA [7]. The figures given were computed over the NIST named curve secp224r1, also known as [45]. It provides equivalent security to a 2048-bit RSA key.

The operations performed during the DTLS handshake are Elliptic Curve Diffie Hellman (ECDH) for key-agreement followed by a two-way authentication via the Elliptic Curve Digital Signature Algorithm (ECDSA) to avoid Man-in-the-Middle attacks.

5.5 Comparison to Related Work

The existing implementation in [35] shows that a DTLS client can be implemented in TinyOS for constrained devices. The use of RSA in the handshake, while being the most widespread public key crypto system in the Internet, basically constitutes the worst-case scenario: RSA keys are very large in size and have high computational demands. Nevertheless, the authors of this book chapter demonstrated that a DTLS handshake with a RSA cipher suite is feasible in WSNs. It was proven that devices are successful in completing such a hand-shake even if packet loss occurs while only requiring a moderate amount of energy for each handshake (under 500 mJ). Handshakes using other public key crypto systems, such as ECC, can therefore be seen as an easy replacement of the basic RSA handshake, because they offer equivalent security at considerably shorter key lengths [8]. However, the authors are not aware of equivalent devices to TPMs that offer secure storage and hardware accelerated computations with ECC keys. Therefore, the implementation for the OPAL mote is superior to ECC based implementations, in that regard, because the authors make use of the Opals TPM for RSA computations and key storage. Although they think that DTLS is a feasible choice for an end-to-end security protocol, there is still room for improvement, which will be described in the next Section. Additionally, DTLS can-not be used to protect routing information and to guard against attacks on a network's availability. Only link layer protocols are able to achieve these goals to a certain degree (cf. Sect. 2.2). Therefore, the authors acknowledge a need for link layer security in WSNs. In use cases like the Internet of Things, link layer and application layer security may complement each other: An application layer security protocol can protect the integrity and confidentiality of application layer messages sent in the local WSN while an additional link layer security protocol can protect the authenticity of the routing information in the local network. In this scenario, only authenticity is needed on the link layer because confidentiality of messages is achieved on the application layer. If messages encrypted with the application layer security protocol leave the local network and enter the Internet they are still protected by encryption, which a link layer protocol cannot achieve alone. Therefore, the presented solution is only compared to other work on application layer and network layer security protocols in WSNs:

Sizzle [17] and SSNAIL [20] both provide SSL or TLS application level security over a reliable transport protocol with 80 bits of security during the handshake through 160-bit ECC keys or 1024-bit RSA keys. The presented DTLS implementation does not require reliable transport, such as TCP, but supports unreliable transport

Table 4 Software ECC over 224-bit prime curve (secp224r1) on OPAL. One of each operation is required for a handshake

	Current (mA)	Computation time (ms)	Energy consumption (mJ)
EC-DH @ 48 MHz	30	387	46.4
ECDSA sign @ 48 MHz	30	432	51.8
ECDSA verify @ 48 MHz	30	795	95.4
Handshake ECC total @ 48 MHz	**30**	**1,614**	**193.6**
EC-DH @ 96 MHz	48	187	35.8
ECDSA sign @ 96 MHz	48	205	39.3
ECDSA verify @ 96 MHz	48	380	72.9
Handshake ECC total @ 96 MHz	**48**	**772**	**92.6**

via UDP. Datagram transport is often preferable in WSNs, because it introduces fewer overheads and is more efficient. Additionally, the authors offer 112-bits of security through 2048-bit RSA keys, which are recommended until the year 2030. 80-bits of security were only recommended until 2010 [39]. Neither of these implementations supports secure storage of private keys in a TPM or similar device. If ECC is used, both, Sizzle and SSNAIL, need about one second to perform a handshake and they need between 6 and 7.5 s when using RSA. It is comparable to the handshakes performance (about 4 s) without the influence of packet loss. However, both implementations require a lot less memory than the presented implementation in this book chapter: Between 2.8 kByte of RAM for Sizzle and 6.3 kByte of RAM for SSNAIL versus the presented solution requirement of 17.2 kByte RAM, which is largely caused by the additional buffer space required for the much larger RSA certificates.

Tiny 3-TLS [46] is different from the presented implementation, because it uses a trusted gateway for trust delegation. In the DTLS implementation, the sensor nodes are independent from a central gateway. The authors of this book chapter cannot directly compare it against performance figures of Tiny 3-TLS. However, if the gateway is fully trusted, the nodes need to perform only symmetric operations, which are much faster than asymmetric operations. If the gateway is only partially trusted, they perform a Diffie-Hellman operation that requires modular exponentiation and may therefore also be resource intensive.

IPsec [47] is independent of the transport layer, because it is located on the network layer. It therefore supports reliable as well as unreliable transport. Raza et al. did not implement the Internet Key Exchange (IKE) protocol, but in-stead deployed the key material prior to their experiments. Their RAM requirements are therefore not comparable to the authors because the handshake, which is a large contributor to memory usage in the DTLS implementation, is not present in theirs. The network overhead of compressed IPsec for payload packets is only 24 Byte, whereas the network overhead of DTLS is currently still up to 64 Byte.

6   Conclusion

This book chapter introduced a standards-based security architecture with two-way authentication for IoT. The DTLS-based solution presented was developed for resource-full devices. Furthermore, a list of important security issues must be kept in mind, when developing an appropriate solution for two-way authentication. The solution for a two-way authentication was described in detail and evaluated in order to justify its applicability for resource-full constrained devices.

In this solution for resource-full devices the authentication is performed during a fully authenticated DTLS handshake and based on an exchange of X.509 certificates containing RSA keys. The extensive evaluation, based on real IoT systems, shows that the proposed architecture (cf. Fig. 1) provides message integrity, confidentiality, and authenticity with affordable energy, end-to-end latency, and memory overhead.

Thus, a DTLS is a feasible security solution for emerging IoT. A fully authenticated handshake with strong security through 2048-bit RSA keys is considered as feasible for sensor nodes equipped with a TPM chip, because a fully authenticated, RSA-based handshake consumes as little as 488 mJ. The memory requirement of fewer than 20 kByte RAM are well below the 48 kByte of memory offered by the used sensor node. Sensor nodes without a TPM chip forego protection against physical tampering, but can still perform a DTLS handshake based on ECC, which could be performed on the chosen platform with little more than 100 mJ of energy consumption.

A similar solution for resource-less constrained devices was implemented supporting also two-way authentication but not working with DTLS and certificates instead using ECC to save resources. This solution achieves the same security goals as the presented DTLS-based solution but is more resource-efficient. For more information it is referred to [8, 26]. Other work has demonstrated techniques to minimize packet headers for similar protocols [21]. For the future, it is planned that these techniques apply to DTLS together with an Authenticated Encryption with the Associated Data (AEAD) mode of operation to achieve a reduction of network overhead.

Acknowledgements The DTLS solution presented was supported partially by the German Federal Ministry of Education and Research: the SODA Project under Grant Agreement No. 01IS09040A and the AutHoNe Project under Grant Agreement No. 01BN070[25]. The standardization activity within IETF was supported partially by FLAMINGO and SmartenIT, funded by the EU FP7 Program under Contract No. FP7-2012-ICT-318488 and No. FP7-2012-ICT317846, respectively.

References

1. Lehong, H., Velosa, A.: Hype cycle for the internet of things. White Paper, Stamford CT, Gartner Inc (2012)
2. European Telecommunications Standards Institute: Machine-to-machine communications (M2M); Smart Metering Use Cases (2010)
3. Leontiadi, I., Efstratiou, C., Mascolo, C., and Crowcroft, J.: SenShare: transforming sensor networks into multi-application sensing infrastructures. In: Proceedings of European Conference on Wireless Sensor Networks, pp. 65–81, Springer, Heidelberg (2012)
4. Shelby, Z., Bormann, C.: 6LoWPAN: The Wireless Embedded Internet. Wiley, United Kingdom (2009)
5. Shelby, Z., Hartke, K., Bormann, C.: The constrained application protocol (CoAP). Req. Comments **7252**, 1–112 (2014)
6. Dawson-Haggerty, S., Tavakoli, A., and Culler, D: Hydro: A hybrid routing protocol for low-power and lossy networks. In: Proceedings of 1st IEEE International Conference on Smart Grid Communications, pp. 268–273 (2010)
7. Kothmayr, T., Schmitt, C., Hu, W., Brünig, M., Carle, G.: DTLS based security and two-way authentication for the internet of things. Ad Hoc Netw. **11**(8), 2710–2723 (2013)
8. Noack, M.: Optimization of two-way authentication protocol in internet of things. Master thesis, University of Zurich, Communication Systems Group, Department of Informatics, Zurich, Switzerland (2014)
9. Bellare, M., Canetti, R., and Krawczyk, H.: Keyed hash functions and message authentication. In: Proceedings of Advances in Cryptology, pp. 1–15 (1996)
10. Karl, H., Willig, A.: Protocols and Architectures for Wireless Sensor Networks. Wiley, England (2007)

11. Miorande, D., Siciari, S., De Pellegrini, F., Chlamtac, I.: Internet of things: vision, applications and research challenges. Ad Hoc Netw. **10**(7), 1497–1516 (2012)
12. Atzori, L., Iera, A., Morabito, G.: The internet of things: a survey. Comput. Netw. **54**(15), 2787–2805 (2010)
13. Bormann, C., Ersue, M., Keranen, A.: Terminology for constrained-node networks. Req. Comments **7228**, 1–17 (2014)
14. Akyildiz, I.F., Su, W., Sankarasubramaniam, Y., Cayirci, E.: Wireless sensor networks: a survey. Comput. Netw. **38**(4), 393–422 (2002)
15. Raymond, D.R., Midkiff, S.F.: Denial-of-service in wireless sensor networks: attacks and defenses. IEEE Pervasive Comput. **7**(1), 74–81 (2008)
16. Luk, M., Mezzour, G., Perrig, A., Gligor, V.: MiniSec: A secure sensor network communication architecture. In: Proceedings of 6th ACM International Conference on Information Processing in Sensor Networks, pp. 470–488 (2007)
17. Gupta, V., Wurm, M., Zhu, Y., Millard, M., Fung, S., Gura, N., Eberle, H., Shantz, S.C.: Sizzle: a standards-based end-to-end security architecture for the embedded internet. Pervasive Mob. Comput. **1**(4), 425–445 (2005)
18. Hu, W., Tan, H., Corke, P., Shih, W.C., Jha, S.: Toward trusted wireless sensorn networks. ACM Trans. Sens. Netw. **7**(1), 5 (2010)
19. Chan, H., Perrig, A., Song, D.: Random key predistribution schemes for sensor networks. In: Proccedings of IEEE Symposium on Security and Privacy, pp. 197–213 (2003)
20. Jung, W., Hong, S., Ha, M., Kim, Y.J., Kim, D.: SSL-based lightweight security of IP-based wireless sensor networks. In: Proceedings of IEEE International Conference on Advanced Information Networking and Applications Workshops, pp. 1112–1117 (2009)
21. Raza, S., Voigt, T., Rödig, U.: 6LoWPAN extension for IPsec. In: Proceedings of Workshop Interconnecting Smart Objects with the Internet, IAB, pp. 1–3 (2011)
22. Raza, S., Voigt, T., and Jutvik, V.: Lightweight IKEv2: a key management solution for both the compressed IPsec and the IEEE 802.15.4 security. In: Proceedings of the IETF Workshop on Smart Object Security, pp. 1–2 (2012)
23. Raza, S., Trabalza, D., Voigt, T.: 6LoWPAN compressed DTLS for CoAP. In: Proceedings of 8th IEEE International Conference on Distributed Computing in Sensor Systems, pp. 287–289 (2012)
24. Winter, T., Thubert, P., Brandt, A., Hui, J., Kelsey, P., Levis, K., Pister, K., Struik, R., Vasseur, J.P., Alexander, R.: RPL: IPv6 routing protocol for low-power and lossy networks. Req. Comments **6550**, 1–157 (2012)
25. Schmitt, C.: Secure data transmission in wireless sensor networks. Ph.D. thesis, Technische Universität München, Institut für Informatik, pp. 1–190 (2013)
26. Schmitt, C., Stiller, B., Noack, M.: Two-way authentication for internet of things. White Paper, IETF ser. ACE Working. Group **14**, 1–19 (2014)
27. Cooper, D., Santesson, S., Farrell, S., Boeyen, S., Housley, R., Polk, W.: Internet X.509 public key infrastructure certificate and certificate revocation list (CRL) profile. Request for Comments, **5280**, pp. 1–151 (2008)
28. Watro, R., Kong, D., Cuti, S., Gardiner, C., Lynn, C., Kruus, P.: TinyPK: securing sensor networks with public key technology. In: Proceedings of 2nd ACM Workshop on Security of AdHoc and Sensor Networks, pp. 59–64 (2004)
29. Modadugu, N., Rescorla, E.: The design and implementation of datagram TLS. In: Proccedings of Network and Distributed System Security Symposium, pp. 1–13 (2004)
30. Ning, P., Liu, A., Wenliang, D.: Mitigating DoS attacks against broadcast authentication in wireless sensor networks. ACM Trans. Sens. Netw. **4**(1), 1–35 (2008). doi:10.1145/1325651. 1325652
31. Schmitt, C., Kothmayr, T., Benjamin, E., Wen, H., Braun, L., Carle, G.: TinyIPFIX: an efficient application protocol for data exchange in cyber physical systems. Comput. Commun. **74**(2), 63–76 (2016)
32. Blake-Wilson, S., Menezes, A.: Authenticated Diffie-Hellman key agreement protocols. In: Proceedings of the Selected Areas in Cryptography, pp. 339–361 (1998)

33. Advantic: CM5000-Datasheet. White Paper, pp. 1–20 (2015). http://www.epssilon.cl/files/EPS5000.pdf
34. Jurdak, R., Klues, K., Kusy, B., Richter, C., Langendoen, K., Brunig, M.: OPAL: a multiradio platform for high throughput wireless sensor networks. IEEE Embed. Syst. Lett. **3**(4), 121–124 (2011)
35. Kothmayr, T.: A security architecture for wireless sensor networks based on DTLS. Master's thesis, Technische Universität München, pp. 1–83 (2011)
36. Atmel: Smart ARM-based flash MCU - Datasheet. White Paper, pp. 1–1163 (2015). www.atmel.com
37. Atmel Corporation: The atmel trusted platform module. White Paper, pp. 1–4 (2007). http://www.atmel.com/images/doc5128.pdf
38. Grossschaedl, J., Tillich, S., Rechberger, C., Hofmann, M., Medwed, M.: Energy evaluation of software implementations of block ciphers under memory constraints. In: Proceedings of Conference on Design, Automation and Test in Europe, pp. 1110–1115 (2007)
39. Barker, E., Barker, W., Burr, W., Polk, W., Smid, M.: Recommendation for key management - Part 1: General (Revised). White Paper, National Institute of Standards and Technology, pp. 1–143 (2007)
40. McGrew, D.A., Viega, J.: The galois/counter mode of operation (GCM). White Paper, National Institute of Standards and Technology, pp. 1–43 (2005)
41. yaSSL: Implementation and performance of AES-NI in CyaSSL embedded SSL. White Paper, pp. 1–14 (2010). http://www.yassl.com/files/whitepapers/whitepaper_883_cyassl_aesni.pdf
42. Liu, A., Ning, P.: TinyECC: a configurable library for elliptic curve cryptography in wireless sensor networks. In: Proceedings of 5th International Conference on Information Processing in Sensor Networks, pp. 245–256 (2008)
43. Kothmayr, T., Schmitt, C.: Tiny pkc implementation. http://www.corinna-schmitt.de/tinypkc.html Check the reference (year, page number)
44. GNU: The GNU general public license version 2. White Paper (1991). http://www.gnu.org/licenses/gpl-2.0.html
45. NIST: Recommended elliptic curves for federal government use. White Paper, pp. 1–43 (1999)
46. Fouladgar, S., Mainaud, B., Masmoudi, K., Afifi, H.: Tiny 3-TLS: a trust delegation protocol for wireless sensor networks. In: Levente, B., Gligor, V.D., Westhoff, D. (eds.) Proceedings of the Third European Conference on Security and Privacy in Ad-Hoc and Sensor Networks, pp. 32–42. Springer, Heidelberg (2006)
47. Raza, S., Chung, T., Duquennoy, S., Dogan, Y., Voigt, T., Rodig, U.: Securing internet of things with lightweight IPsec. SICS Technical report, 1–27 (2011)

Mobility Support and Service Discovery for Industrial Process Monitoring

Tuan Dang, Pascale Minet, Patrick Bellot, Christophe Mozzati and Erwan Livolant

Abstract A service-oriented architecture designed for the use of wireless sensors networks (WSN) in industrial applications such as the operation and maintenance of industrial installations is presented in this chapter. This architecture comprises the OCARI wireless sensor network and the OPC-UA/ROSA middleware, as well as the KASEM predictive maintenance system. In contrast to well-known communication stacks for wireless sensor networks such as ZigBee and WirelessHart, OCARI has been designed to support mobility of sensor nodes. Furthermore, the OPC-UA/ROSA middleware provides service discovery to enable the interconnection of the WSN with the Internet of Things (IoT). This architecture targets various industrial applications such as process monitoring, pollutant detection, monitoring of fuel storage area, fire detection in temporary worksites, health monitoring of people working in hazardous conditions, etc.

T. Dang (✉)
EDF R&D Lab, STEP Department, 6, Quai Watier - BP 49, 78401 Chatou Cedex, France
e-mail: tuan.dang@edf.fr

P. Minet · E. Livolant
Inria, 2 Rue Simone Iff, 75012 Paris, France
e-mail: pascale.minet@inria.fr

E. Livolant
e-mail: erwan.livolant@inria.fr

P. Bellot
LTCI, Télécom ParisTech, Université Paris-Saclay, 75013 Paris, France
e-mail: bellot@telecom-paristech.fr

C. Mozzati
Predict, 19 Avenue de la Foret de Haye, 54519 Vandoeuvre-les-nancy, France
e-mail: christophe.mozzati@predict.fr

© Springer International Publishing AG 2017
D.P. Acharjya and M. Kalaiselvi Geetha (eds.), *Internet of Things: Novel Advances and Envisioned Applications*, Studies in Big Data 25, DOI 10.1007/978-3-319-53472-5_3

1 Introduction

Implementing new sensors for industrial process monitoring raises several challenges: interfaces to the qualified sensors (already stocked in warehouses) that often use a 4–20 mA current loop, wiring between the sensors and an acquisition system (sometimes a programmable logic controller) that induces significant costs of preparation including costly modifications to existing equipment and the need for technical expertise to configure communication systems, etc. The use of wireless sensor networks is an alternative to limiting or even eliminating these drawbacks. Another challenge concerns the optimization of radio frequency (RF) spectrum in industrial environment. Indeed, the spectrum is a limited resource that users must share correctly to avoid interference and to ensure electromagnetic compatibility (EMC) with existing instrumentation and control devices.

The choice of RF technologies that may coexist and be scalable is therefore important for functional safety and quality of service requirements. The embedded or added RF transmitter in wireless instrumentation should satisfy the following requirements:

1. The communication software should allow seamless deployment of wireless sensor networks by any operational team that has no telecom specialist. This corresponds to the so-called "plug-and-play" characteristic of the wireless sensor network. Any effort required for the manual configuration or reconfiguration of the network during operation may greatly reduce the interest of wireless sensor networks.
2. The ability to reuse existing and qualified sensors that use a 4–20 mA current loop. These represent a large base in continuous process industries. This means that the RF transmitter should have an interface allowing it to be integrated in a 4–20 mA current loop.
3. The ability to interface seamlessly with existing monitoring information systems for predictive maintenance of equipment. This means that using industrial standards such as OPC-UA [1] for exposing measurement services is essential.
4. The total cost of ownership (i.e. the total cost of acquisition and operating costs) should be as low as possible. It must be much lower than that of wired instrumentation.

These requirements are essential for the penetration of wireless sensor networks in industrial applications. This chapter presents a service-oriented architecture that allows seamless and easy integration of wireless sensor networks in an information system for predictive maintenance. This architecture is illustrated through the example of a wireless sensor network using the OCARI protocol [2, 3] and the OPC-UA/ROSA [4, 5] middleware based on OPC-UA.

Industrial end-users are generally very reluctant to accept proprietary technology concerning hardware and software platforms. Indeed, many industrial process installations have a long life cycle; a 30-year life cycle is very common. However, in the electronics industry, the life cycle of hardware is much shorter than in an industrial installation. Partial renovation of instrumentation and control systems is a common solution to make the installation durable. Proprietary solutions present risks in terms of durability if the manufacturer does not durably maintain its hardware and software line of products. Partial renovations are easier to carry out if standard and widely available platforms are used. For this reason, standard and widely available platforms for the development of the OCARI wireless sensor network and the OPC-UA/ROSA middleware have been chosen.

The chapter is structured as follows: Sect. 2 presents a brief state-of-the-art regarding WSNs and middleware used in industrial applications. Section 3 recalls the main characteristics of OCARI and then focuses on mobility support. Section 4 presents the middleware OPC-UA/ROSA and defines its interconnection with the OCARI WSN. In Sect. 5, the proposed service-oriented architecture is described. In our experience, this architecture is a key point for a successful development of wireless sensor networks supporting industrial process monitoring. It integrates the KASEM predictive maintenance system. In Sect. 6, the easy integration of the OCARI protocol stack is illustrated for various types of sensor nodes. The open development environment of the whole architecture is presented. We also show how to make the best use of OCARI by tuning its parameters to meet the requirements of a given application. Finally, an example of an industrial application using this architecture is given for a use case of fire detection and for the detection of a leaking valve. Future work and development perspectives are given in Sect. 7.

2 State of the Art

This state-of-the-art is organized in two parts. The first is related to wireless sensor networks, and particularly the OCARI WSN. The second part presents industrial middleware with a focus on OPC-UA.

2.1 Wireless Sensor Networks

With regard to wireless networks, there is a plethora of solutions and standards, as depicted in Fig. 1. Different types of wireless networks can be distinguished according to their geographic extent, ranging from WPAN/WBAN for the smallest to the WMAN/WWAN for the largest.

Taking into account the robustness requirements of industrial applications leads us to select mesh technologies where routing is able to adapt to topology changes, which excludes SIGFOX and LoRa. SIGFOX and LoRa are two long range technologies based on a star topology providing a radio range of a few kilometers and also low throughputs of a few kilobits/s. LoRa uses the frequency band of 868 MHz, and this band is constrained by ETSI to meet a duty cycle of 1%. Furthermore, the use of this band outside Europe is an open issue under current regulations. LoRa

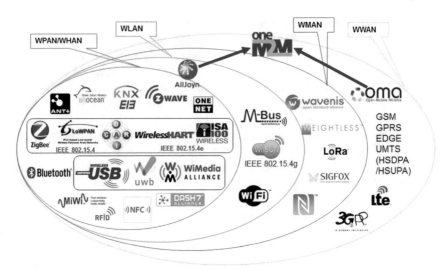

Fig. 1 A plethora of wireless networks

provides a bidirectional communication. SIGFOX works in the 70 MHz frequency band and provides a throughput lower than LoRa. In addition, the communication is unidirectional from the sensor node to the base station. To improve delay and throughput, a channel is assigned to each sensor node. However, the number of available channels is usually much smaller than the number of sensor nodes, and the medium access contention must be solved. Neither LoRa nor SIGFOX provides a complete and standardized protocol stack.

In contrast, in meshed networks, a node is able to communicate with several neighbors. This possibility is used by the routing protocol to build a path from this sensor node to the sink, this path usually being multi-hop. The failure of a node or a link can be recovered using other links. In addition, nodes can transmit with a power less than the maximum and as a consequence they can save energy and prolong the lifetime of their battery. Furthermore, the spatial reuse of frequencies is made easier. The physical layer of the IEEE 802.15.4 standard constitutes a consensus for the development of wireless sensor networks. Freescale, TI, NXP, Microchip, ST Micro, and Atmel propose chips and development platforms based on this standard whose robustness, evaluated by its bit error rate is good.

Solutions based on meshed topologies but different from OCARI is further discussed below. WirelessHART [6] is the result of cooperation between major industrial companies such as Emerson and ABB. This solution is standardized CEI 62591:2010. It is mainly implemented by Dust Networks or Linear Technology, for temperature, pressure, flow, and level meters. ISA100 [7] is relevant for the control command industry. This solution is standardized CEI/PAS 62734:2012. It supports IPv6 by means of 6LoWPAN. There are some products available, provided by Yokogawa.

Both WirelessHART and ISA100.11a have brought extensions to the MAC layer of IEEE 802.15.4 with the objective of making the medium access more deterministic.

ZigBee [8] supports a wide range of applicative profiles, public or private, though not standardized in this latter case. It uses the AODV routing protocol that does not take into account the residual energy of nodes. In addition, mobility may induce poor performances. 6LoWPAN [9] has been developed to promote the paradigm of the Internet of Things, where any object will communicate in IPv6. However, to take into account the strong limitations of wireless sensor networks such as a MAC payload reduced to 117 bytes and a throughput of 250 kbits/s in the IEEE 802.15.4 standard, some compression has been done to optimize the performance. 6LoWPAN is used with the RPL ('ripple') routing protocol [10]. This routing builds a directed acyclic routing graph, where each node selects its preferred parent and the neighbor that provides the best metric value. RPL allows the use of several metrics and assumes an implicit agreement on the choice of the metric used. Up to now, the number of implementations has remained limited and there is no industrial support. The use of RPL with mobile nodes raises some issues regarding the instability of routes and the medium access contention created by control messages. Finally, Table 1 summarizes these different solutions.

Table 1 Comparison of mesh-based wireless solutions

	ZigBee	WirelessHART	6LoWPAN-CoAP	OCARI
Routing	Routing table is updated only after the current path is broken: some performance issues	Static routing graph computed at the initialization	RPL routing based on directed acyclic graphs	EOLSR (Extension of RFC 3626) dynamic and Proactive routing
Energy	Residual energy of nodes is ignored	Residual energy of nodes is ignored	Residual energy of nodes may be considered according to the metric chosen	Residual energy of nodes taken into account in route building
Topology Scalability	Tree, star and mesh Maximum depth limited to 15 in ZigBee Pro	Star and mesh 10 nodes per network	Tree, star and mesh	Tree, star and mesh 50 nodes per network (cluster)
Medium access	IEEE 802.15.4 CSMA-CA without GTS, the PAN coordinator selects the channel used, without determinism	IEEE 802.15.4 with extension TDMA and channel hopping, some determinism (but difficult to predict because of channel hopping)	IEEE 802.15.4 standard, without determinism	MaCARI (IEEE 802.15.4 improved with beacon cascading), determinism and spatial spectrum reuse, energy efficiency (OSERENA node coloring)

In conclusion, the solutions presented have all or some of the following disadvantages. These disadvantages are briefly discussed below.

1. *Static or centralized building of routes for WirelessHART and ISA100*: The routing graph must be predefined. It is then needed to train the users and the team in charge of the WSN deployment. Reactivity to a topology change after a node or link failure may be slow and the latency will be high. As a consequence, mobility of nodes cannot be supported. The WSN deployed lacks flexibility. It is not easy to replace a sensor node by another one.
2. *Centralized medium access without spatial reuse of the spectrum*: WirelessHART and ISA100 are based on a TDMA which assigns one different slot to each node. This rapidly raises performance problems: latency increases proportionally to the number of nodes and the throughput granted to each node is inversely proportional to the number of nodes. Scalability is not achieved.
3. *Energy is not taken into account when building routes*: WirelessHART, ISA100 and ZigBee do not consider energy, be it the energy consumed during the transfer of a message from its source to its final destination, or the residual energy of intermediate nodes. This accelerates the failure of these nodes by battery depletion.

2.2 Industrial Middleware

A middleware stands between the operating system and the components of a distributed application. In the case of industrial middleware, the applications are usually linked to embedded and realtime data acquisition and computing. Middleware in Information Technology are based on service oriented architecture (SOA), and the industry is seeking better performance in terms of efficiency, footprint, operational cost, etc. For instance, the adaptability of SOA for sensor networks is discussed in [11, 12].

A middleware must ensure reliability and performance while handling interoperability. Although this is the case for CORBA (Common Object Request Broker Architecture [13]), it is not the case for all the DCOM/COM based middleware (such as OPC, the forerunner of OPC-UA) which are Windows based. These two kinds of middleware represent about 50% of the industrial middleware currently in use. Because of recent attacks on industrial systems and terrorist threats, the middleware must ensure some kind of authentication and security, for example signature of message. For instance, coming from the Internet of Things [14]), the M2M protocol MQTT [15], implementing a publish-subscribe scheme, supports authentication and even TLS-security at the price of a large network overhead [16]. As MQTT claims to be simple and lightweight, these features do not seem to be used with current device technology. It also seems that protocols coming from the IoT, such as CoAP [17] and MQTT, cannot be used as industrial middleware for the present since there are so many industrial constraints.

Another constraint is the fact that the middleware should provide its clients with meta data (such as types for instance). This is useful for building an application starting from scratch and being able to show data from a sensor for instance. Through its address space and its specialised nodes, OPC-UA [18], is able to represent and provide to its clients with a lot of meta data about the devices to which the server is connected and the values that it delivers elsewhere. The middleware, DDS [19], provided by the OMG data distribution service, does not support meta information as OPC-UA does and, in another area, does not seem suitable for resource-constrained systems, and OPC-UA may have different footprints depending on the type of implementation.

There exist a lot of middleware and tools, either for industrial control command or for controlling wireless sensor networks [20–22]. Most of these have interesting features, but their technology readiness level (TRL) is not high enough. To be adopted by industry, a middleware should be standardized so that two companies building two different devices will be able to communicate through the standard. Of course, DDS is a standard but it has not been widely adopted in industrial applications such as supervision and control in which OPC and OPC-UA have become well established. The situation may be the same for IoTivity, OneM2M/SmartM2M and All Seen which tend to target mass market applications. OPC-UA has a communication stack which is standardized and any company can build a device implementing this stack. This device can be put in any OPC-UA environment, even if it was built by another company.

3   The OCARI Wireless Sensor Network

In this section, the main principles of OCARI are presented together with mobility support. OCARI is a low power consumption and low bandwidth RF communication protocol for industrial wireless sensor networks. It is designed to meet:

1. The possibility of having routers or intermediate sensor nodes operating with a limited power source (e.g. battery). This requires optimizing power consumption by avoiding collisions of data messages (e.g. measurements) by scheduling node activity, and by switching nodes to sleep mode when they have nothing to send.
2. Self-configuration and "self-healing" of the network when deploying sensors and when link or router failure occurs.
3. The spatial reuse of the RF spectrum to shorten the time needed for data gathering.
4. The use of hardware compliant with the IEEE 802.15.4 standard.

OCARI can be distinguished from ZigBee and WirelessHART by the following two characteristics such as a medium access method that combines CSMA/CA, and an optimized TDMA based on a three-hop node coloring with spatial reuse of time slots. This coloring algorithm, called OSERENA [23] also provides an activity scheduling mechanism that helps to reduce interference and thus optimizes the node's power consumption. A routing strategy called EOLSR [24] (Energy efficient

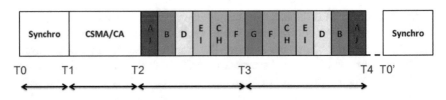

Fig. 2 The OCARI global cycle

OLSR [25]) which is power efficient and dynamically maintains mesh topology of the network is employed. This takes into account the node's residual energy. The operating cycle of OCARI is depicted in Fig. 2 where the activity is organized into five periods [T0–T1], [T1–T2], [T2–T3], [T3–T4], and [T4–T0'].

1. The period [T0–T1] deals with multi-hop deterministic synchronization of nodes using cascaded beacons [26]. The network controller, also called the CPAN, broadcasts the beacon including the ordered sequence according to which this beacon is forwarded in the network. As a consequence, the beacon is propagated without any collision over the whole network. This beacon contains useful information such as the length of the different periods in the OCARI cycle.
2. The period [T1–T2] deals with transmission of data and signaling messages in competition using CSMA/CA.
3. The period [T2–T3] deals with upstream transmission of data messages originated from sensor nodes without collision in colored slots.
4. The period [T3–T4] deals with downstream transmission of data messages originating from the sink are disseminated to sensor nodes without collision in colored slots. During this period, the network coordinator transmits any command message to sensor nodes. Such a message is routed downstream the routing tree used for data messages.
5. The period [T4–T0'] deals with sleep activity. In this period, all nodes switch to a deep sleep mode, saving more energy from microprocessors and memory.

The spatial reuse of time slots is illustrated in Fig. 3 where node A and node J share the same time slot as they are four hops away from each other. It is the same for node E and node I or for node C and node H. Thus, in Fig. 2, the red, green and blue slots are assigned to two transmitters.

A wireless sensor node seamlessly joins the OCARI network. After discovering its neighbors, a route is created toward the sink in charge of collecting data measured by this sensor node. This node is assigned a color. It is then able to transmit its data toward the sink in its colored slot, a time slot in the [T2–T3] period, without any collision. Upon detecting a link or node failure, routes are automatically updated to avoid the broken link or the failed node.

OCARI is power efficient, since each sensor node knows exactly when it must be awake to receive data from its children and send them to its parent in the routing tree. Each node sleeps when it has no messages to send or receive. As a consequence, there is no energy wasted due to collisions or idle listening. OCARI also supports

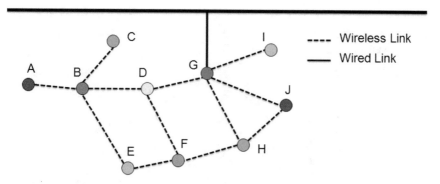

Fig. 3 A topology illustrating the spatial reuse of time slots

mobile nodes. Any message sent by any mobile node M in [T1–T2] is forwarded by the static node N that is heard by M with the highest signal strength (i.e. the highest RSSI). This message is then propagated in the collision-free [T2–T3] period. When M moves, the static node selected, changes accordingly. This mobility support is detailed in Sect. 3.1.

3.1 Mobility Support

Although major industrial applications in WSNs are static, there will be an increasing number of mobile nodes. These mobile nodes can be human beings wearing sensors to monitor their health in hazardous environments or mobile and autonomous robots performing tedious tasks. In this context, mobile nodes move at a pedestrian speed. The number of mobile nodes is assumed to have the same order of magnitude as the number of static nodes. The basic assumption is that, whatever its location in the OCARI network, any mobile node is able to have a bidirectional communication with a static OCARI node. In addition, a very limited number of mobile nodes may act as mobile sinks, gathering data from some mobile nodes.

3.1.1 Mobility Support Principles

To briefly summarize mobility support, the static nodes build a wireless backbone to which mobile nodes attach themselves. The static node to which a mobile node attaches varies when the mobile node moves. The role of mobility support is to allow these changes without inducing a large overhead. This is a necessary condition for a non degradation of the quality of service provided to static nodes.

In the association procedure, any node N that wants to be associated in the network, selects the static node that it receives with the strongest signal. The node N receives

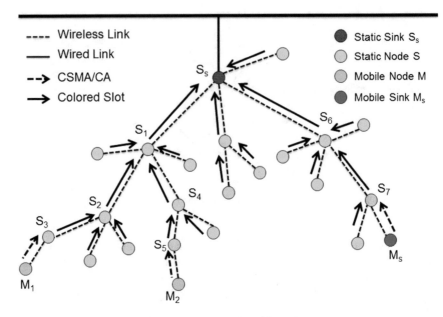

Fig. 4 Upstream communication in a topology with mobile nodes

its permanent short address. There is no change in this procedure, except that node N, static or mobile, must declare its type: static, mobile or mobile sink. The attachment procedure is run exclusively by mobile nodes. After its association, a mobile node selects the static node that it is associated with as its attachment parent. Then, each time the quality of the wireless link with its attachment parent goes below a given threshold for received beacons, the mobile node selects another attachment parent. The attachment procedure is an implicit one i.e., it does not require any message exchanges between the mobile node and its attachment parent.

For an upstream communication, two cases are distinguished, depending on the source of the communication. When the source is a static node, data are sent in colored slots. When the source is a mobile node, data are sent in CSMA/CA for the first hop, and in colored slots for the next hops. Figure 4 illustrates upstream communication from mobile nodes in an example of OCARI topology. The sequence of data transmissions in the different parts of the OCARI cycle are illustrated in Fig. 5.

For a downstream communication, several cases are also distinguished according to the final destination of the message. When the message is intended for all the nodes, it is broadcasted first by the sink and then by static nodes in colored slots.

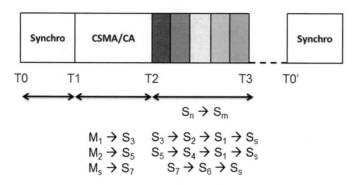

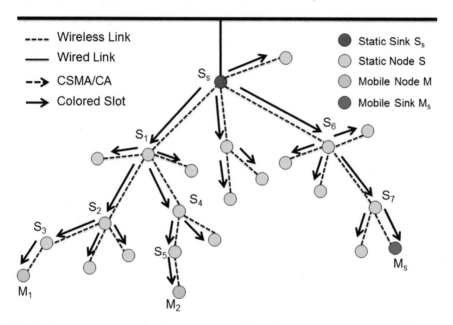

Fig. 5 Upstream communication from mobile nodes in an OCARI cycle

Fig. 6 Downstream communication in a topology with mobile nodes

The application layer of each node receives this message. A mobile node listens to the downstream colored slots until it correctly receives the messages included in a colored slot. At the end of first colored slot of reception, the mobile node can sleep. A static node listens only to the downstream colored slot of its parent. Figure 6 illustrates upstream communication from mobile nodes in an example of OCARI topology. The sequence of data transmissions in the different parts of the OCARI cycle are depicted in Fig. 7.

When the message is intended for a group of nodes, it is broadcasted first by the sink and then by static nodes in colored slots like the previous case. But, unlike the

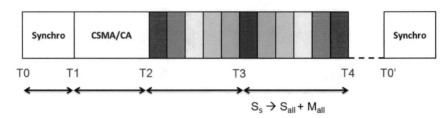

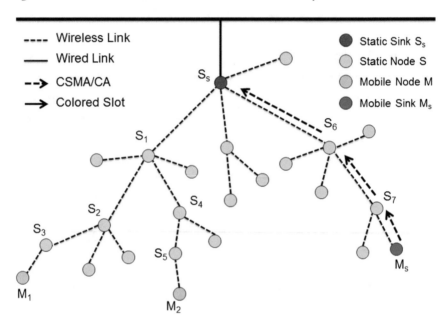

Fig. 7 Downstream communication to mobile nodes in an OCARI cycle

Fig. 8 Route record from the mobile sink

previous case, only the nodes belonging to the multicast group deliver this message to their application layer. When the message is intended for a single node that is not a mobile sink, it is broadcasted first by the sink and then by static nodes in colored slots like the previous case. Only the final destination node delivers this message to its application layer. When the message is intended for a mobile sink, the message is sent first by the sink and then by static nodes in unicast in the CSMA/CA period. This message is routed to the mobile sink according to the route included in the message (i.e. source routing). The route to the mobile sink is known by the static sink because it has previously received a message generated by the mobile sink and recording its route.

Figure 8 depicts the route recording functionality initiated by the mobile sink M_S and applied by all static nodes on the upstream route toward the static sink S_S.

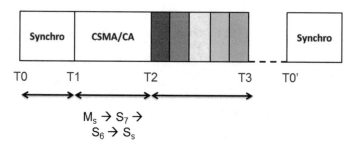

Fig. 9 Route record toward the static sink in CSMA/CA

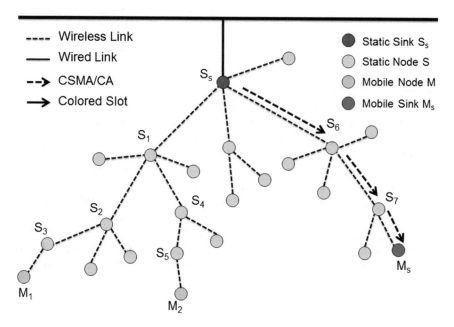

Fig. 10 Source routing from the static sink to a mobile sink

Figure 9 depicts the transmission of the route recording message in the CSMA/CA part of the OCARI cycle.

Figure 10 depicts the source routing functionality initiated by the static sink S_S and applied by all static nodes on the downstream route toward the mobile sink M_S. Figure 11 depicts the transmission of the route recording in the CSMA/CA part of the OCARI cycle.

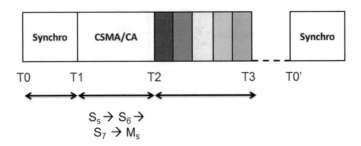

Fig. 11 Source routing toward the mobile sink in CSMA/CA

3.1.2 Mobile Nodes

Mobile nodes are kept very simple, to limit the overhead they could induce each time they move. At the MAC layer, mobile nodes implement the attachment procedure. The attachment parent is changed each time the quality of the wireless link with its attachment parent goes below a given threshold. The new parent is selected among the static nodes whose beacon has been received with a sufficiently strong signal strength. Notice that the mobile nodes never send beacons. They transmit their data in the CSMA/CA period. They are awake in the downstream colored slots to be able to receive data, broadcast to all nodes or unicast to a mobile node.

At the Network layer, mobile nodes implement neither the EOLSR routing protocol that is in charge of neighborhood discovery by means of "Hello" messages, and route building toward the sink by means of STC messages, nor the OSERENA coloring algorithm that assigns a color to each static node. They only use a default router that is attached to their parent, to which they forward any messages they send.

3.1.3 Mobile Sink

Mobile sinks are also simple, they are similar to mobile nodes, except that they generate a route recording message each time they change their attachment parent. This overhead seems acceptable, since on one hand the number of mobile sinks is very limited and on other hand they are the final destination of messages generated by the static sink. More precisely at the MAC layer, mobile sinks implement the attachment procedure as any mobile node. Notice that the mobile sinks never send beacons. They transmit their data in the CSMA/CA period. They are awake in the downstream colored slots to be able to receive data and to broadcast to all nodes. They may also receive data from the sink in the CSMA/CA period. At the network layer, mobile nodes implement neither EOLSR, nor OSERENA. Like any mobile node, they only use a default router which is their attachment parent. Each time this attachment parent is changed, a mobile sink generates a route record message recording its route to the sink.

3.1.4 Static Nodes

To support mobility, static nodes have to implement two additional simple functionalities at the network layer. These are route recording and source routing. In route recording, a static node receiving a message with the route records flag set records its address in the message before forwarding it. In source routing, a static node receiving a message with the source routing flag set routes this message according to the route recorded in the message.

4 The OPC-UA/ROSA Middleware

OPC-UA is an object-oriented middleware platform. It represents the state-of-the-art middleware technology for the industrial automation and control command domain. It is one of the middleware technologies implementing the IEC 61850 standard and has been sometimes elected as the most prominent smart grid integration technologies. The version of OPC-UA presented here is named OPC-UA/ROSA. It is made of a C++ implementation of the middleware OPC-UA coupled with an overlay network named ROSA, which stands for robust overlay network with self adaptive technology [4, 5]. The main purpose of OPC-UA/ROSA is to provide a standard and low cost interface to the world of OCARI wireless sensor networks.

The abstract specification of the OPC-UA may be seen as an M2M communication protocol. The primary aim is to create an open standard for industrial control systems with very fast protocols such as UA-TCP that allows permanent communication between two machines, for instance a client and a server operating a sensor. The server operating sensor encodes the data into binary using the UA-Binary format. Additionally, OPC-UA supports service oriented architecture (SOA) through the HTTP-SOAP protocol, mainly for information management using web Services. It also supports security through a TLS-like security named UA secure conversation for the UA-TCP protocol and through WS secure conversation for the HTTP-SOAP protocol.

This standard is multi-platform and claims to be interoperable. Actually, many implementations of OPC-UA are carried out in many languages. All these OPC-UA implementations must be interoperable provided they support the OPC certification. The OPC-UA stack is in charge of serialization, signing, encoding, and transport of packets. The implementation described in this chapter is under the LGPL v3 licence, and is available from the Git repository of the CONNEXION project (www. git.cluster-connexion.fr/git/opc-rosa.git). It runs under Windows, Linux 64-bit and 32-bit, OSX and Raspbian.

The OPC-UA standards (IEC 62541) are written in 13 parts. The implementation of all OPC-UA standards would lead to a very large footprint. However if taken minimum, for instance a server operating a sensor device with the minimum set in the address space, will get a very low level footprint module and is named as profile. Some implementations are supported by a device such as the Raspberry or even a less powerful device [27]. The services that an OPC-UA implementation must propose are the secure channel services set and the session services set. These services are necessary to open a communication between a server and a client. If the implementation is not concerned with security, these parts may be reduced. The other categories of services that may be implemented are listed below.

1. Discovery service set that allows a client to discover services.
2. Data access services set that allows a client to read and write attributes of all objects in the address space, including the values of the variables.
3. Historical data access services set that allows the client to manage the different values of OPC-DA according their production time.
4. Alarms and events services set to subscribe to alarms via a publish subscribe scheme.
5. The call services set which allows methods in the server to be defined and to call them with parameters from the client.
6. Finally, there exist the browse services set and view services set to explore and manage the address space of a server.

Only the UA-TCP protocol has been implemented with the serialization of data using UA-Binary. Different types of security such as UA-None (no security), UA-Sign (signing transferred data) and UA-SignAndEncrypt (encrypting them too) have been implemented. The OPC-UA service discovery set is not implemented because the ROSA overlay network implements another contextual and distributed services discovery. The historical data access and the view service set are not necessary for the use cases considered inside the CONNEXION project (www.cluster-connexion.fr). Otherwise, most of the OPC-UA specifications have been implemented, sometimes with a few restrictions. The address space of a server can be filled when the server starts running by using XML files. All security schemes and all services implemented can be disabled at the compilation time of the software so that it is possible to resize the resulting software as desired.

4.1 The ROSA Overlay Network

ROSA is an overlay network (ON). Like most ONs, ROSA must be run on every computer that is part of the industrial information system. It relies on the notion of a lump: a lump is a set of nodes of the network such that two nodes are interconnected by the network (a clique in graph theory). There exists a maximum number of lumps that a node can belong to, and there exists a maximum number of nodes which can belong to a lump. The density of a lump is the minimal number of failures on the network links that are necessary to break the lump. It can be computed using the min cut or max flow algorithm. If the number of link failures is less than the density, then every member of a lump is still connected with all the other nodes. ROSA may be seen as an entanglement of lumps and is always trying to locally increase the density of lumps by joining a node to a lump, splitting lumps and so on. This is the automated and decentralised management of ROSA.

The purpose of ROSA was to automatically recover from failures of the underlay network. If the usual routing path between two nodes is interrupted, ROSA automatically finds another route waiting for an IGP protocol to recover the routing in the network. In this case, the IGP protocol such as RIP, IGRP, OSPF, etc. will repair the underlay routing very fast and so ROSA in this case, is useless. Still research work is going on this part of ROSA to make it faster. Nevertheless, ROSA, like many ONs, proposes a distributed hash table (DHT) where the key value pairs are stored in all the nodes of a lump. In this way, the DHT of ROSA becomes resilient and a node failure does not affect the DHT.

4.2   The Link Between OPC-UA and ROSA

The link between OPC-UA and ROSA is illustrated in Fig. 12. If a server or a client wants to store, search or delete a service, it uses ROSA. If a client or a server is unable to send a message, it uses ROSA in its transport layer. However it has not been completely discussed.

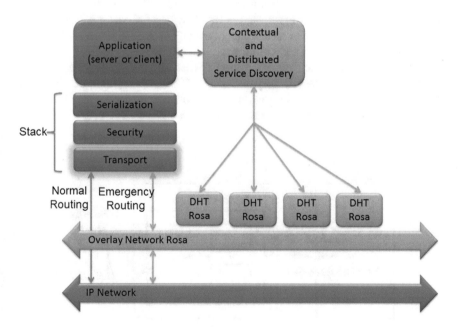

Fig. 12 Relation between OPC-UA and ROSA

4.3   The Link Between OCARI and OPC-UA

The OPC-UA server acts as a middleware between the OCARI wireless sensor network and any application that needs to get data from any wireless sensor node or
to send data to any wireless sensor node. Typically in OPC-UA, this is done via
variables in the address space of the server. An OCARI wireless sensor corresponds
to a variable in the address space of the server. The reading of this variable corresponds to getting the value of the sensor, it is depicted in Fig. 13. The link between
the variables in the server and the OCARI network is done using a dynamic library
in Windows, a shared object in Linux and a dylib in OSX shortly written as DLL.

When a new sensor arrives in the OCARI network, the DLL creates a variable
with a name from OCARI and it pushes the service description in service discovery.
Each order of the OCARI network, such as the order to change a valve position
as described in Sect. 5, is triggered by calling the appropriate technique with the
appropriate parameters. The goal of the technique is to send the order to the OCARI
network and to return a status code. The integration of the OCARI wireless sensor
network in the industrial information system is made by a gateway as depicted in
Fig. 14. From the hardware point of view, the gateway between OCARI and OPC-
UA/ROSA is made up of the OCARI module of the wireless network controller
(CPAN) and a Raspberry Pi B+, connected by a serial link on a USB cable. From the
software point of view, the CPAN transmits its data to the Raspberry in a serial way
according to a pivot frame format. On the Raspberry, an OCARI driver working in
Raspbian makes it possible to record data in the OPC-UA/ROSA server and makes
these data available to any OPC-UA client.

In this architecture, each wireless sensor is seen as a service that publishes its
functional capacities in the directory of the OPC-UA/ROSA middleware. This middleware is in charge of the orchestration of the publish or subscribe paradigm to the
available services authorized by the standard mechanism of OPC-UA.

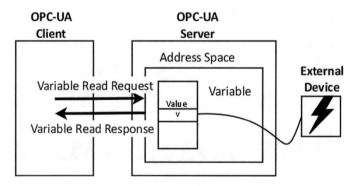

Fig. 13   OPC-UA as a middleware between the OCARI wireless sensors network and any application

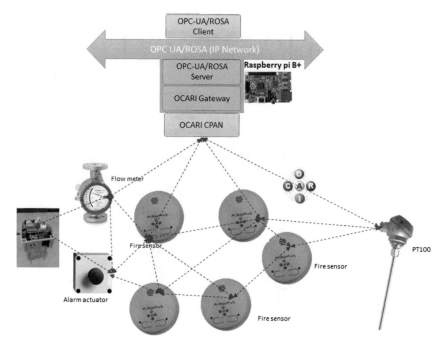

Fig. 14 Gateway between OCARI and OPC-UA/ROSA

5 The KASEM Predictive Maintenance System

A service-oriented architecture is a key aspect in accelerating the deployment of wireless sensor network applications. The deployment of a wireless sensor network application in an industrial installation often faces several challenges.

1. Existing industrial information systems are generally designed to provide functionalities such as Supervisory Control and Data Acquisition (SCADA). In such an architecture, most of the instrumentation must be specified in advance. Any insertion of new instrumentation generally requires a reconfiguration of the SCADA system, and thus makes the system unavailable.
2. The development of a specific gateway for each specific wireless instrumentation. This may be costly.
3. The development of a specific application to deal with each specific measurement format, etc.

To meet these challenges, this chapter presents an SOA depicted as in Fig. 15. It presents instrumentation as a service and provides several standard services that may be easily combined.

Data publication service: The Data publication service consists in the OCARI wireless sensor network and the OPC ROSA middleware, both of which have been described above.

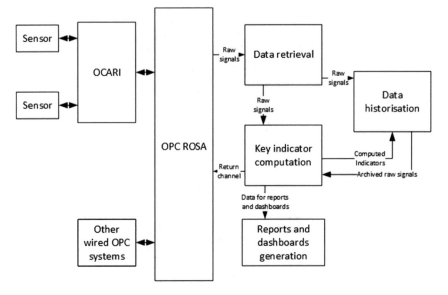

Fig. 15 An example of service architecture based on OCARI and OPC-UA/ROSA

Computer aided operation and maintenance: The KASEM software platform (www.predict.fr/en/products/kasem/) from PREDICT is used. KASEM [28, 29] is a predictive maintenance tool that computes health assessment indicators and detects drifts or abnormal behavior. To do so, it needs the widest possible range of sensor data, hence the interest in using a wireless sensor network to complement the existing wired acquisition. Additionally, KASEM also provides several services listed as below.

1. *Data retrieval service*: This service retrieves time series from sensors using different communication channels. It can be real-time stream such as OPC or data files such as CSV files. In this case, an OPC-UA/ROSA client is developed and integrated inside the KASEM acquisition architecture. This client subscribes to the variable tags exposed by the OPC-UA/ROSA server. The embedded OCARI gateway updates the tags each time a new value is read from the wireless sensor network. These new values are retrieved every 1 s in KASEM. A data visualization tool is embedded inside the KASEM web portal, as shown in Fig. 16. This data retrieval service is used with several OCARI-equipped sensors such as temperature probes, level meters and flowmeters.
2. *Key indicator computation*: KASEM has a built-in computation service that can be triggered in two ways such as real time way and differed time way. Each time a data point arrives from the data retrieval service, KASEM can compute an indicator based on the current value of several variables. In differed time, using the same code, it is possible to recompute a batch of data from a large time period. The built in computation engine uses C# code. KASEM provides a library of mathematical models dedicated to health assessment computation and drift detection.

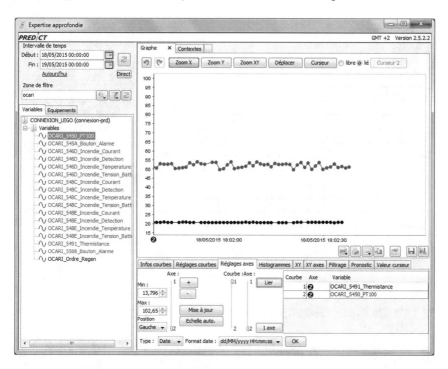

Fig. 16 OCARI Sensor acquisition as seen in KASEM

It is also possible to create custom code directly in the KASEM Web portal using an embedded code editor. A typical computation includes signal inputs, parameters and signal outputs. Computation can be arranged in sequences, to divide a complex algorithm into several simple steps. In this CONNEXION project, a simple example is the monitoring of flowrate. The signals of several wireless flowmeters positioned on different pipes are acquired. By comparing them, and correlating them with temperature and pressure data, problems at various parts of the process can be detected. Figure 17 depicts the whole process of industrial early detection monitoring.

Return channel communication: A real time algorithm can send back signals to the OCARI sensors using the OPC-UA/ROSA bus. To do so, OPC-UA methods on sensors objects can be called by the KASEM Computation engine. Each time a method is called on a node, the message is broadcasted to the OCARI network and propagated to the nodes. There are several use cases for this, such as action on a wireless node (e.g. change valve position), parameter configuration (e.g. increase sample rate because of a degraded situation) and enable or disable a sensor. The whole chain is depicted in Fig. 18.

Event generation: Another output of the algorithm is event generation. The KASEM platform provides a four step workflow service for each event to be

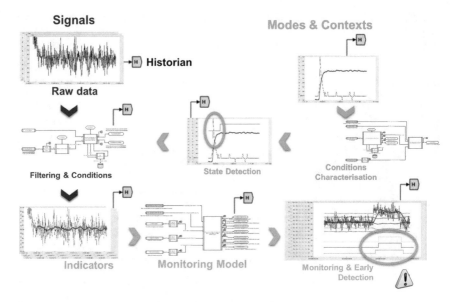

Fig. 17 Industrial early detection monitoring

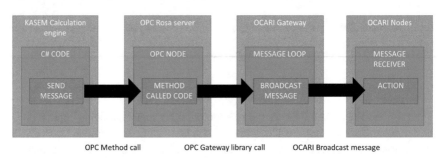

Fig. 18 Return channel mechanism

processed by a human operator. In this workflow, a comment can be attached to the event, diagnostic tools can help the operator to identify the root cause of the event. Further work orders can be issued to a CMMS software to fix the causes. The information filled is reused as feedback each time a similar event is triggered. An example of event generation by KASEM is illustrated in Fig. 19.

Report and dashboard generation: Reports and dashboards provide a way to vizualize synthetic indicators about the process health. Typically, they integrate computed indicators that are aggregation of process health status. It can be displayed in the KASEM web portal or output as files (PDF, Word, HTML). Figure 20 illustrates an example of a dashboard generated by KASEM.

Data historization: The KASEM knowledge base stores every data point received from the data retrieval service and every output from the calculation service. KASEM uses an SQL server database as a back-end for this knowledge base. Each data point

Mobility Support and Service Discovery for Industrial Process Monitoring 79

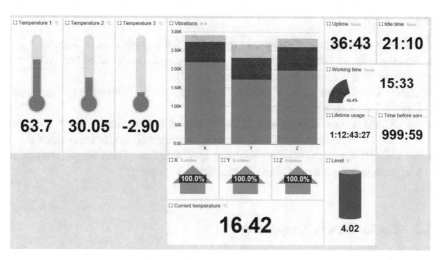

Fig. 19 Example of event generation by KASEM

Fig. 20 Example of a dashboard generated by KASEM

is stored with its timestamp, its status (good, bad value, error, overflow, etc.) and its value. The raw data is also used by the computation engine for batch indicator calculation.

6 Implementation and Use Cases

The architecture described in this chapter has been implemented in several use cases. For example, the use case of temperature and air flow monitoring in the process of decommissioning a power plant in which hundreds of OCARI wireless sensors are involved. Another use case concerns temperature and humidity monitoring in data center in which dust must be avoided. A use case in which mobility support is required to monitor personal exposure to radiation where maintenance workers are moving inside a nuclear power plant. The hardware and software that are used to develop the proposed solution are described.

6.1   The OCARI Development Platform

OCARI protocol is currently implemented on Atmel SAM3S4A platform (an ARM Cortex M3 microcontroller) coupled with the Atmel AT86RF233 RF transceiver. Such hardware may be bought off-the-shelf from Dresden Elektronik GmbH under reference deRFsam3-23T09-3 (www.dresden-elektronik.de) or from Adwave under reference ADWRF24-LRS (www.adwave.fr). The software development platforms include the followings.

- A source code management system using Git (www.gitlab.com/iiot/ocari).
- Atmel studio, an integrated development environment under Microsoft Windows that may be downloaded from the Atmel website.
- The GNU cross-compiler for ARM processor "arm-none-eabi-gcc" and "openocd" for developers who prefer Linux.
- An OCARI frame dissector plugin for Wireshark (www.wireshark.org) available from the Git repository at OCARI website (www.gitlab.com/iiot/wireshark).
- An OCARI topology network analyzer available from Inria forge (www.scm.gforge.inria.fr/anonscm/git/contiki-hiper/contiki-hiper.git).

6.2   Sensor Nodes Supported

Many types of sensor nodes are supported by OCARI, as detailed below. More precisely, we have developed a small printed circuit board (PCB) that may be easily integrated with any existing sensors using a 4–20 mA current loop, depicted in Fig. 21.

Several types of sensor such as the temperature sensor (PT100), the flowmeters (e.g. Krohne H250 and Krohne OPTIFLUX) and the guided level radar meter (e.g. Krohne OPTIFLEX 2200) integrate the OCARI protocol by using the above PCB for process monitoring as shown in Fig. 22. With these sensor nodes, two industrial applications have been developed, based on two different use cases.

1. Intelligent fire detection with different detectors. The use of different types of fire detectors in a mesh topology allows us to discard false alarms and to identify the nature of fire.

Fig. 21 PCB (5 cm x 3 cm) for integration with existing sensors using a 4–20 mA current loop

Fig. 22 Examples of sensors for process monitoring using the OCARI protocol

2. Detection of a leaking valve using various types of flowmeters, level sensors and PT100s sensor in the cooling circuit in a maintenance context. For example, level sensors are used to detect the loss of water in a storage basin belonging to the cooling circuit. PT100s are used to detect an abnormal change in temperature in a cooling circuit which means that there is a leak somewhere between two measurement points upstream and downstream of the suspected valve.

6.3 OCARI Tuning

This section show how to tune the OCARI parameters in order to meet the requirements of a given application. An application is characterized by requirements in terms of latency, denoted *Latency*, robustness and energy consumption. The WSN topology is defined by the number and positions of sensor nodes deployed in the area considered. The question is how to make the best use of the OCARI network for this application and considering this network topology. The following assumptions are defined.

1. Assumption A1: The slot size enables each node to aggregate the messages received from its children with its own message.
2. Assumption A2: The message containing the own data of any node is available at the beginning of the OCARI cycle if this node has no children, and when the messages of the children are received otherwise.

In the absence of message loss and buffer overflow, it can be proved that:

1. With assumption A1, the data sent by any sensor node in an OCARI cycle are delivered to the sink in the same cycle.
2. With assumptions A1 and A2, the data sent by any sensor node are delivered to the sink in a time that is less than or equal to the duration of the colored period.
3. with assumption A1 only, the data sent by any sensor node are delivered to the sink in a time that is less than or equal to the duration of the OCARI cycle plus the duration of the colored period.

Hence, the tuning of OCARI parameters should meet the following in Eq. (1).

$$OCARI\ cycle + colored\ period \leq Latency \tag{1}$$

Before computing the different periods of the OCARI cycle, some notation is introduced. Let nbs be the number of static nodes; nbm be the number of mobile nodes; nbc be the number of colors, and bs be the size of a beacon slot. Let $CSMAs$ be the CSMA/CA time assigned per static node. The default value is 20 ms corresponding to the transmission time needed by four messages of maximum size (127 bytes of physical payload). Let $CSMAm$ be the CSMA/CA time assigned per mobile node. The value should be set to the number of messages sent per mobile node per cycle times the time needed to transmit a frame of maximum duration. Let $COLORs$ be the duration of a colored slot. The default value is set to 32 ms, allowing the transmission of 5 messages of maximum size. The duration of the synchronization period is given by the following Eq. (2).

$$[T0 - T1] = nbs \cdot bs \tag{2}$$

The duration of the CSMA/CA period is given by Eq. (3).

$$[T1 - T2] = nbs \cdot CSMAs + nbm \cdot CSMAm \tag{3}$$

The duration of the colored period is given by Eq. (4).

$$[T2 - T3] = (nbc - 1) \cdot COLORs \tag{4}$$

The duration of the downstream colored period is given by Eq. (5).

$$[T3 - T4] = nbc \cdot COLORs \tag{5}$$

The length of the inactivity period can be computed, taking into account the latency required by the application as expressed in Eq. 1. The duration of the inactivity period is given by Eq. (6).

$$[T3 - T0'] = Latency - [T0 - T1] - [T1 - T2] - 2[T2 - T3] \tag{6}$$

7 Conclusion

This chapter presents a service-oriented architecture that is used for the development of applications such as condition based maintenance using the KASEM system. The implementation of such an architecture is illustrated through the use of the OCARI wireless sensor network and the OPC-UA/ROSA middleware. The solution proposed to support mobility in the OCARI network is simple and limits the overhead induced by mobile nodes. This mobility support is designed to be efficient in its use of resources such as bandwidth, energy, memory. The properties of energy efficiency, determinism, latency and robustness provided by OCARI to static wireless sensor nodes are ensured. In the absence of mobile nodes, the OCARI network behaves exactly as without mobility support and exhibits exactly the same performances. Similarly, the overhead induced by mobile sinks is only incurred if mobile sinks are present. As data gathering by the static sink is the most important objective of the OCARI network from the application point of view, its performances are not altered by mobility support. Data gathering by a mobile sink is a new functionality provided by mobility support. Since it has a lesser degree of importance for the application, it is provided with a quality of service that may be lower.

Two main obstacles limit the widespread use of the IoT in industrial environments. The first is the unavailability of tools able to report information about the quality of wireless links to assist the deployment of WSNs. The development of such tools is the next step of our work. The second obstacle is the lack of statistical information collected by the WSN protocol about any wireless link, in order to improve the robustness of the WSN in the case of mobility. The design of such protocol extensions is also in our research workplan. The proposed implementation of the OPC-UA middleware, namely OPC-UA/ROSA, serves as a normalized gateway to the WSN OCARI. Starting from this gateway, operating as an OPC-UA server, IoT industrial applications can be developed with OPC-UA clients that are able to consult values of sensors and to operate actuators. These applications have now been working for months. Moreover, this scheme is secured using X.509 V3 certificates and TLS. Additionally, OPC-UA server can run on a nano computer Raspberry Pi showing that the footprint of OPC-UA can range from very large (where everything is implemented) to very small (where only what is necessary is implemented). The next step is to make this footprint even smaller so that it can be run on a smartphone.

The use of wireless sensors with the KASEM predictive maintenance platform has enabled it to provide new capabilities in computing process health management (PHM) indicators. By correlating signals coming from wired and wireless instrumentation, a more precise health status can be computed for the system. OPC-UA/ROSA middleware allows KASEM to use multiple heterogeneous data sources without the need for specific development. The next step is to apply OPC-UA/ROSA middleware and OCARI WSN to a larger scale and study their applicability in a health monitoring and prognostic assessment in a fleet context.

Acknowledgements We would like to thank our partners in the CONNEXION project (www.cluster-connexion.fr).

References

1. Wagner, D.: OPC-UA specification, Part 1 to Part 12. Technical Report on OPC-UA Foundation, pp. 1–7 (2009)
2. Khaldoun, A.A., Gerard, C., Alexandre, G., Erwan, L., Saoucene, M., Pascale, M., Michel, M., Joseph, R., Thierry, V., van den Adrien, B.: Cross-layering in an industrial wireless sensor network: case study of OCARI. J. Netw. **4**(6), 411–420 (2009)
3. Tuan, D., Pascale, M., Erwan, L.: OCARI: a wireless sensor network for industrial environments. ERCIM News **15**(101), 3455–3467 (2015)
4. Tuan, D., Dragutin B., Patrick, B.: Routing in OPC-UA with rosa overlay network. In: Proceedings of Workshops on the Move to Meaningful Internet Systems, pp. 86–90 (2014)
5. Loïc, B., Nguyen, P., Patrick, B.: Robust overlay network with self-adaptive topology: Protocol description. In: Proceedings of IEEE International Conference on Research, Innovation and Vision for the Future in Computing and Communication Technologies, pp. 154–160 (2008)
6. Haung, K..: HART communication protocol specification. Tech. Rep. HART Commun. Found. **12**(4), 129–137 (2008)
7. Isa, K..: ISA-100 wireless systems for industrial automation: process control and related applications. Tech. Rep. Int. Soc. Autom. **12**(4), 112–122 (2009)
8. Zig Bee Alliance: Zig Bee Specification, pp. 16–29 (2008)
9. Montenegro, G., Kushalnagar, N., Hui, J., and Culler, D.: Transmission of IPv6 packets over IEEE 802.15.4 networks. Req. Comments **4944**, 1–30 (2007)
10. Winter, T., Thubert, P., Brandt, A., Hui, J., Kelsey, R., Levis, P., Pister, K., Struik, R., Vasseur, J.P., Alexander, R.: RPL: IPv6 routing protocol for low-power and lossy networks. Req. Comments **6550**, 1–157 (2012)
11. Scholz, A., Gaponova, I., Sommer, S., Kemper, A., Knoll, A., Buckl, C., Heuer, J., Schmitt, A.: Epsilon SOA - service oriented architectures adapted for embedded networks. In: Proceedings of 7th IEEE International Conference on Industrial Informatics, pp. 599–605 (2009)
12. Mohamed, N., Al-Jaroodi, J.: A survey on service-oriented middleware for wireless sensor networks. Serv. Oriented Comput. Appl. **5**(2), 71–85 (2001)
13. Orfali, R., Harkey, D., Edwards, J.: The Essential Client/Server Survival Guide, 2nd edn. pp. 1–9. Wiley, New York (1996)
14. Atzori, L., Iera, A., Morabito, G.: The internet of things: a survey. Comput. Netw. **54**(15), 2787–2805 (2010)
15. Edwards, J.: MQTT Specification. Technical Report on OASIS Standard, pp. 1–8 (2014)

16. Gaponova, I.: MQTT and the NIST Cybersecurity Framework. Technical Report on OASIS Standard, pp. 1–5 (2014)
17. Shelby, Z., Hartke, K., Bormann, C.: The constrained application protocol (CoAP). Req. Comments **7252**, 12–19 (2014)
18. OPC-UA Foundation: OPC UA Specification, part 3: Address Space Model, Version 1.01. pp. 1–9 (2009)
19. Kirov, G., Stoyanov, V., Lazarov, B.: Abstract model of an object-oriented layer for distributed systems based on the DDS standard. In: Proceedings of the 12th International Conference on Computer Systems and Technologies, pp. 75–81 (2011)
20. Perera, C., Zaslavsky, A., Christen, P., Georgakopoulos, D.: Context aware computing for the internet of things: a survey. IEEE Commun. Surv. Tutor. **16**(1), 414–454 (2014)
21. Henricksen, K., Robinson, R.: A survey of middleware for sensor networks: state-of-the-art and future directions. In: Proceedings of the International Workshop on Middleware for Sensor Networks, pp. 60–65 (2006)
22. Linthicum, D.S.: Enterprise application integration, pp. 1–47. Addison-Wesley Longman Ltd, UK (2000)
23. Ichrak, A., Pascale, M., Cedric, A.: OSERENA: a coloring algorithm optimized for dense wireless networks. Int. J. Netw. Distrib. Comput. **1**(1), 9–24 (2013)
24. Saoucene, M., Pascale, M.: EOLSR: an energy efficient routing protocol in wireless ad hoc and sensor networks. J. Int. Netw. **9**(4), 389–408 (2008)
25. Clausen, T., Jacquet, P.: Optimized link state routing protocol (OLSR). Req. Comments **3626**, 19–24 (2003)
26. Gerard, C., Erwan, L., Alexandre, G., Adrien-van-den, B., Michel, M., Thierry, V.: Specifications and evaluation of a MAC protocol for a LP-WPAN. Ad Hoc Sens. Wirel. Netw. **7**(1/2), 69–89 (2009)
27. Succic, S.: Optimizing OPC UA middleware performance for energy automation applications. In: Proceedings of IEEE International Conerrence on Energy, pp. 1570–1575 (2014)
28. Voisin, A., Medina-Oliva, G., Monnin, M., Léger, Jean-Baptiste, Iung, B.: Health monitoring and prognostic assessment in a fleet context. In: Proceedings of Annual Conference of the Society For Machinery Failure Prevention Technology, pp. 566–571 (2014)
29. Voisin, A., Medina-Oliva, G., Monnin, M., Léger, Jean-Baptiste, Iung, B.: Fault diagnosis system based on ontology for fleet case reused. In: Ebrahimipour, V., Yacout, S. (eds.) Ontology Modeling in Physical Asset Integrity Management, pp. 133–169. Springer International Publishing, Switzerland (2015)

MAC Protocols in Body Area Network-A Survey

Manish Kumar and Mayank Dave

Abstract Body area networks represent the natural union between connectivity and miniaturization. Formally, it is defined as a system of devices in close proximity to a persons body that cooperate for the benefit of the user. These networks are appealing to the researchers due to their wide range of application areas such as remote health monitoring, fitness monitoring, triage, biosensors etc. However, typical properties of body area network bring the necessity to achieve an efficient medium access protocol in terms of power consumption and delay. Various medium access protocols with different objectives were proposed for wireless sensor network by many researchers. This chapter throws light on some features that make body area networks different from wireless sensor networks. A comparative study of various medium access protocols for body area networks is also presented.

1 Introduction

Advancement of wireless technology gave new directions to healthcare applications. Removing wires and cables enables patients to benefits of increased mobility. Main concern in current senario is the increase of chronic diseases such as cardiovascular, hypertension and diabetes among aged people. This makes it necessary to find out a wireless technology to support remote patient monitoring in an unobtrusive, reliable, and cost effective manner [1]. Since the time, size of sensors and actuators have decreased marginally, these devices can be used for monitoring patients and controlling therapeutic functions remotely. Ultra low power wireless connectivity among devices placed in, on, and around the human body is seen as a key technology enabling unprecedented portability for monitoring physiological signs in the

M. Kumar (✉)
PEC University of Technology, Chandigarh, India
e-mail: manishkamboj3@gmail.com

M. Dave
National Institute of Technology, Kurukshetra, India
e-mail: mdave67@gmail.com

© Springer International Publishing AG 2017   87
D.P. Acharjya and M. Kalaiselvi Geetha (eds.), *Internet of Things:*
Novel Advances and Envisioned Applications, Studies in Big Data 25,
DOI 10.1007/978-3-319-53472-5_4

hospitals, at home, and on the move. Body area networks (BAN) is one such emerging technology that has the potential to significantly improve health care delivery, diagnostic monitoring, disease tracking, and related medical procedures. A BAN is defined formally as a system of devices in close proximity to a persons body that cooperate for the benefit of the user. Additionally, it can act as a local area network. It can also be termed as "on the user for the user". Thus, BANs would offer continuous, real-time health diagnosis in day-to-day life and in sports activities, emergency, and military services [2].

BAN is an emerging area of wireless communication. But, the level of information that can be provided by the sensors is limited. In addition, energy resources that can be capable of powering such sensors are not available. The technology is still in its primitive stage and it is being widely researched [3]. This technology is expected to be a breakthrough invention in healthcare once adopted. It will lead to the concepts like telemedicine and mhealth. According to IEEE 802.15, a BAN is defined as "a communication standard optimized for low power devices and operation on, in or around the human body (but not limited to humans) to serve a variety of applications including medical, consumer electronics, personal entertainment, and other". The development of BAN technology was started around 1995 by considering wireless personal area network (WPAN) technologies for communications on, near, and around the human body. Later around 2001, this application of WPAN has been named as BAN to represent the communications on, in, and near the body only [4].

BAN can also be considered as an advancement of cyborg. It is otherwise known as cybernetic organism and deals with both biological and artificial enhancements e.g. electronics, mechanical or robotics. It came into existence in 1960 when M Clynes and S. Nathan Kline used it in an article about the advantages of self regulating human machine systems in outer space. This term is often applied to an organism that has enhanced abilities due to technology. In 1998, Kevin Warwick had implanted in his arm, a radio frequency identity (RFID). As a result, he could turn on lights by snapping his fingers. He used his mind to control a robotic hand, and successfully transferred his thoughts across the Atlantic and clenched a mechanical fist. The English scientist says that it is time for us to overcome our human limitations. In future, a man will be able to use more than his five senses, as the implants of chips in human brain will stretch our ways of communicating with people and objects.

The basic concept of BAN is the fusion of two main ideas such as a set of mobile, compact units that enable transfer of vital parameters between the patients location and the clinic or the doctor in charge. The vital signs data flow passes a chain of BAN modules from each sensor to a main body station, which consolidates the data streams of all sensor modules attached. It transmits the data to home base station, from where they can be forwarded to the concerned authority via telephone line or Internet.

Many projects has been carried out in this direction. Wear-a-BAN is a project that deals with unobtrusive wearable human to machine wireless interface to investigate and demonstrate ultra low power BAN technologies for enabling unobtrusive human to machine interfaces (HMI) [5]. It enabled major technological breakthroughs in the areas of ultra low power radio system-on-chips (SoC). This is also useful in textile oriented system in package (SiP) platforms for miniature wearable antennas, wireless sensor electronics, and digital signal processing [6]. Another project named home based empowered living for Parkinsons disease patients (HELP) developed a comprehensive system that is capable to administer drug therapy [7]. It proposes a process that can be either continuous or on-demand basis, to manage disease progression and to mitigate Parkinsons disease. Similarly, the MobiHealth consortium unites 14 partners from hospitals and medical service providers, universities, mobile network operators, mobile application service providers, mobile infrastructure, and hardware suppliers [8]. It focuses on full mobility of patients while undergoing health monitoring. The patients wear a lightweight monitoring system, the MobiHealth BAN. This is customized to their individual health needs. Therefore, a patient who requires monitoring for short or long periods does not have to stay in hospital for monitoring. Similarly, another project named AYUSHMAN has a vision to provide a dependable, secure, real-time automated health monitoring.

For a successful implementation of BAN, a standard model is required that can address both medical and non-medical applications. IEEE established a task group called IEEE 802.15.6 for the standardization of BAN in November 2007 [9]. At present, IEEE 802.15.6 standard defines three physical layers such as narrow band (NB), ultra wideband (UWB), and human body communications (HBC) layers [10]. The selection of each physical layers depend on the application requirements. On the top of it, the standard defines a sophisticated MAC protocol that controls access to the channel. Mainstream adoption of BAN is still over the horizon as engineers and researchers work to overcome challenges involving interoperability, sensor design constraints that deals with power and complexity, privacy, security etc. [11]. Once these issues are tackled, BAN can revolutionize healthcare with concepts like telemedicine and mHealth to become real. This can also potentially allow for revolutionary uses in communications, security and sports.

An infrastructure of BAN is depicted in Fig. 1. In this infrastructure, various sensors for measuring vital signals are placed on human and the data is collectively forwarded to the personal sever. The personal server can also act as environmental sensor that can measure temperature, humidity of the surrounding and can pass on the data collectively via Internet to outside world. This data can be sent to the doctor and can be copied to medical server for maintaining the report of the patient. In addition, in case of emergency direct call can be made to ambulance.

Fig. 1 Infrastructure of body area network

2   Communications in Body Area Network

Communication in BAN can be broadly classified into two categories depending
upon the range such as on body communication and off body communication. On
body communication is confined in the range of 2 m around the body in which either
the various body sensors communicate with each other or body sensors communicate with the personal server (PS). One subclass of this communication consists of
implanted sensors and are used for in-body communication, in which implanted nodes
communicate among themselves. Due to a different body hindrance and extremely
power conservation nature of in-body nodes, this type of communication plays vital
role in BAN. Off body communication establish the communication between PS and
access points (APs) where APs are fixed in the infrastructure or of adhoc in nature.
The following Fig. 2 represents how data flow in a typical BAN. A brief discussion
on stepwise flow is presented below [12].

Black dots placed on the human body represents the sensors. These sensor can
be of any type like blood pressure, ECG etc. These sensors send data continuously
like one that is sending the heart rate or after a fixed interval of time like sugar
level in blood. These sensors remain in direct contact with a PS that can be patient
mobile phone or PDA. They pass the collected data to the server. All the data is
being collected on PS. One side, PS communicates with the sensors on the body
and at other side, it communicates with the outside world through Internet. The
information collected in first step is transferred to the medical server (MS). In case
of emergency, it can directly make a call to emergency service [13]. This provision is
very helpful in saving the life of patient who needs constant supervision. Physician
retrieve the data from the MS and after analysis recommendation can be sent. These
can be some prescription of exercise or medical dose. Depending upon the type

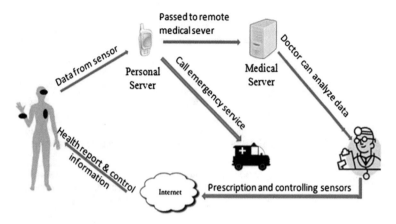

Fig. 2 Flow of data in body area network

of sensors deployed, physician can directly control the actuator also. For example, if insulin pump is deployed in the body of the patient then doctor can use it to control the level of insulin in the patient body depending upon the sugar level. The prescription is sent back to the PS through Internet to patient and patient can read the medical report and can act accordingly. Wireless sensor network (WSN) that has its roots in micro-electro-mechanical systems (MEMS) technology. Wireless communications and digital electronics have enabled the development of low-cost, low power, multifunctional sensor nodes [14]. At first glance, BAN and WSN seems largely same but Table 1 highlights some parameters that distinguish BAN completely from WSN.

3 Challenges and Research Issues in Body Area Network

Body area networks has various challenges. This section highlights various challenges and open research issues pertaining to BAN. These challenges need to be tackled carefully to make BAN successful [15].

1. *Energy efficiency*: Batteries available these days work well for handheld electronics, but their capacity is limited in BAN context. The need to replace or recharge batteries frequently makes BAN less desirable. Nodes should be able to take energy from ambient sources, such as sunlight or vibration, which will make attractive solution to energy woes. BAN sensor nodes must be small and have batteries that last about days to years, depending on the application. The size requirement of nodes for BAN limits the size of the batteries that will power the nodes. Recharging batteries with harvested energy could not only extend battery life, but also simplify BAN use. The energy efficiency is a research challenge that can be handled by making the sleep cycle of the BAN nodes more effective.

Table 1 Comparative analysis of wireless sensor networks and body area networks

Description	Wireless Sensor Network (WSN)	Body Area Network (BAN)
Scale	Wide area coverage (in km)	Limited to human body (in centimeters/a few meters)
Number of nodes	Huge for covering large area	Limited number of application specific nodes
Accuracy	Compensated by redundancy so large number of nodes are used	Accurate measurement are required by each node
Size	Small size are preferable but not important	For in-body sensor size should be smallest possible, on-body node should also be of small size
Node job	All nodes perform a dedicated task	Nodes are designed so that they can perform multiple tasks
Mobility	Nodes are mostly stationary and no group movement of tasks	Nodes on/in the patient move in a group and in same direction
Security	Lower	Higher to protect patient data
Data	Loss can be compensated by redundant data	Not tolerable and some mechanism is required to maintain QoS
Relay	Any node can act as relay	In-body nodes should be avoid to act as relay because of high energy conservation
Power source	Solar power, wind power	Body heat or motion
Biocompatibility	Need not to be considered	Must be there for in-body nodes and on-body nodes
Node replacement	Easy to perform	In-body nodes are difficult to replace
Topology	Static	Dynamic due to patient movement
Wireless technology	Bluetooth, ZigBee, GPRS, WLAN	Some low power technology is required

In addition, node can be placed at such position so that they can get energy from the available resources [16].

2. *Security*: In a typical BAN, there can be security breach at various levels. Starting from the PS, it is important to avoid disclosure and overhearing of medical information. Overhearing of data in BAN may cause severe damage to the patient. The data might be used for some illegal purposes. The confidentiality can be achieved by encrypting the data with a secret key that is known only to PS and MS. Data being received by the PS or MS should be sent by trusted sensor nodes.

Adversary may try to send some false signals to PS that may try to disrupt the normal functionality of the BAN. In addition, if two or more BANs are available then PS should be able to differentiate between the two. This can be achieved by using a shared secret key within a BAN [15, 17].

3. *Location privacy*: One crucial requirement of BAN is location privacy since such network broadcasts highly personal information. It is required to protect this data using available data encryption algorithms. Even though the data within a BAN is encrypted, it is still possible to track the location of the individual by tracking the unique hardware addresses associated with the gadgets. Thus, attacker could pick up the signals of a BAN from a distance using ultra sensitive antennas.

4. *In and on body communication challenges*: As the application field of BAN is vast, the type of communication and hence the radio communication challenges can be divided into two main categories [17]:

   - *In-Body Communication challenges*: Body nodes are implanted under the human skin and have extremely critical power requirement. These type of nodes are self triggered i.e., if the value of certain parameter exceeds a predefined threshold, they reliably send emergency data in least possible time to the hub. Therefore, in-body MAC should accommodate emergency traffic along with ongoing normal traffic. The in-body traffic is categorized as continuous and non-continuous traffic. In continuous, once the nodes receive a wakeup signal from coordinator will keep on sending the data. Non-continuous traffic provides only a snapshot for monthly health check up of a patient. Additionally, in-body nodes are considered as the propagation channel inside the body and the antenna is located near the body. The transmitter power must be limited by the safety guidelines to human exposure to electromagnetic radiations. Also, path loss of signal is high because of propagation environment [8].
   - *On-body communication challenges*: On body nodes are worn on the body and they create a communication system among themselves. Here mostly a mesh network configuration is made and they use industrial, scientific, and medical (ISM) band for data transmission. Here both transmitter and receiver antennas are located very close to the body and are strongly influenced by the presence of the body. Carrier sense multiple access with collision avoidance based MAC techniques can be used, unlike in-body communication where nodes cannot perform clear channel assessment (CCA). The communication in this case makes a star topology.

5. *Medium access control issues*: In BAN, a short range communication network is formed between the nodes. There are many challenges in designing MAC layer for such a network. IEEE 802.15.6 draft does not constitute a complete MAC protocol as they outline only basic elements that must be used to create a BAN such as packet formats and message exchange protocols. The draft leaves many unanswered questions related to retransmission, relays, out of contention, non-contention etc. Finding issues in MAC are important because it can help in making smart decisions in terms of efficiency, reliability and throughput of healthcare system. Further issues of *MAC* are discussed in next subsections [1].

- *Dynamic allocation of slots*: In case of contention free access, packet transfer between sensor and hub can cause collision. It leads to loss of packets. Therefore, a time slot that could have been used productively by some different node is wasted. It can be avoided if near future of the channel can be predicted based on the current state of the channel. For example, if a recent attempt of transmission has failed, then the next slot should be used for retransmission or should be allocated to a different node. This is another open research issue. Further, reliability can be increased if BAN uses variable slot allocation scheme by determining the future state of the network. It looks similar to opportunistic scheduling of cellular networks. But, this traditional approach is not applicable here, as it requires slave nodes to be continuously available for communication that contradicts the energy saving characteristic of BAN.
- *Retransmission scheduling*: As BAN in healthcare demands for close to 100 percent success rate that cannot be achieved without retransmission of failed packets. This retransmission has to be considered with extra energy and delivery deadlines. Also, retransmission depends on the state of the channel [18]. Therefore, efficient methods are to be devised for retransmission like allocating a retransmission slot at the end of each superframe.
- *Relay nodes*: Small outages can be managed by using retransmission but if a node is out of the range of hub for time periods larger than the packet delivery deadline then relay nodes must be used. The best example is when people are sleeping, they move less and so a typical node can have significantly large outage period. Such cases have no alternative other than to route the bad links using a relay node. The node that is to be used as the relay must remain awake to listen for packets that are to be forwarded to hub from the sensor nodes. As the nodes in BAN are energy critical, so relays should be used as efficiently as possible. Also, it is critical in deciding nodes that are to be used as relay nodes [10].
- *Transmit power*: Constantly transmitting at the maximum allowed power of 1mW will increase reliability. But it will result in very high energy consumption. So there is a need to improve both energy efficiency and reliability by controlling the transmit power for each node in the network. Sample and hold transmit power scheme can be used effectively for controlling the transmit power. In such cases, a transmitting node estimates channel reduction from the last packet it has received and then adjusts its transmit power accordingly.

4   IEEE Sandard for Body Area Network

At first glance, most of the existing standards like bluetooth and wireless personal area network (IEEE 802.15.4) seem appropriate for BAN. Table 2 represents a comparative analysis of IEEE BAN standard with other IEEE 802.15 standards. However, when most of the challenges discussed above are taken into account they fail to meet

Table 2 Comparison of BAN standard with existing standards

Description	Other IEEE 802.15.X standards	BAN IEEE 802.15.6 standard
MAC layer protocols	15.3 and 15.4 medium access protocols can be used	Single scalable MAC with reliable delivery is required
Power consumption	Low	Extremely low as the medium here is human body for in-body nodes
Power source	Conventional power source	Conventional or with body energy
Frequency band	ISM or UWB	Medical authorities approved communication bands which are different in different countries

the requirements. In November 2007 IEEE 802.15.6 working group was created to draft physical and MAC layer standards for BANs to address the challenges is [19].

Most important feature of IEEE 802.15.6 BAN is flexibility. It offers multiple physical layer modes of operation: narrowband, ultra-wideband and human body communication. Other great feature is the flexibility by offering a number of different access modes. Time division multiple access (TDMA) is combined with polling which provide the facility of giving exclusive time slots to transmit or receive information [20]. Carrier sense multiple access (CSMA) also allows many ways to combine the different access modes together. To increase the energy efficiency each device need not to implement all of the access mechanisms, it can choose the one that is suitable for its operation.

5 MAC Techniques for Reliable and Efficient Body Area Network

MAC layer plays a significant role in determining lifetime of network by controlling main sources of energy waste like collision, idle listening, polling messages and control packet overhead. Collision occurs when two or more nodes transmit data packet at the same time. This may cause retransmission of the packet that consumes extra energy. Idle listening is the phenomenon in which a node listens to idle channel to receive or transmit data. Overhearing is to get data that are not meant for that node. In BAN there are different physiological signals that are quite variable because of variable data rate and data generation. Some signal appears to be periodic but may turn into non-periodic based on a patient condition. For example, if a patient develops an abnormal heart condition then the heartbeat sensor could generate a non-periodic signal. A MAC protocol for BAN should be able to handle these variations efficiently. IEEE 802.15.6 standard do not contribute to a complete MAC protocol. It just specifies a MAC sub layer in support of several physical layers [21]. Here some of the existing protocols are discussed with the advantages and disadvantages of each.

5.1 Scheduled Time Division Multiple Access MAC Protocol

Scheduled time division multiple access divides the channel access into different time slots. There is no contention for the channel resulting in a deterministic delay and no packet loss. Time division multiple access (TDMA) is a scheduled multiple access technique where transmission of packets are managed in the form of time frames and slots. A time slot can be seen as a dedicated transmission resource used to carry patient data with minimum or no overhead [20]. A central control unit (CCU) allocate slots on a permanent basis to each sensor nodes. These sensor nodes transmit information from it to the CCU. A TDMA based MAC protocol is suitable for a small BAN with a limited numbers of sensors generating data at a fixed rate and transmitting fixed block of data. In TDMA, a sensor node does not need to listen to the transmission channels all the time, hence energy consumptions for the communication circuit is minimized. This scheme is energy efficient where transmitter circuit of a sensor node is only activated in the specified slot. Figure 3 shows a basic TDMA frame divided into different time slots.

TDMA based BAN has two major resource allocation problem. There is no support for non-periodic data and TDMA is not scalable. It is a centralized protocol that requires change in the central controller setup to accommodate any extra sensor in the BAN. It could be the main drawback for a patient monitoring system because number of sensor nodes could vary depending on the condition of a patient. For example, if medical staff wants to add more monitoring devices on a patient body that is under supervision then it will be necessary to modify the design of the TDMA time frame from the main controller. Therefore, this type of scheme is useful for patient monitoring system having a fixed number of nodes.

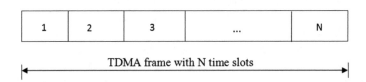

Fig. 3 TDMA frame and time slots

5.2 *Polling Based MAC Protocol*

Polling based MAC protocol is based on master slave relationship for transmitting data. In polling network, CCU schedules all transmissions in the network thus avoiding any contention. As shown in Fig. 4, a polling network is flexible than a fixed assignment based TDMA network. It is better because, traffic sources of different data rates can be accommodated in this protocol. As shown in the figure, CCU will send polling messages to all the nodes which in response to the poll can either transmit a message they have or can send a negative acknowledgment (NACK) if the node has no data to transmit. CCU can also implant an acknowledgment (ACK) for the previous data to improve the reliability [22]. The period of polling of sensor nodes by CCU vary according to the type of sensors.

Figure 5 shows the polling cycle for a round robin scheme where nodes are polled in a sequential manner. At the end of cycle, an additional polling slot is kept to keep the scalability feature of the protocol. If a new node want to join the network it has to send a message to CCU. Also, to leave the network, a node has to inform the CCU. This protocol has two main advantages such as scalability and transmission delay. It is scalable because of last slot in the cycle and transmission delay is predictable. Whenever a new sensor node want to join the network, it sends a JOIN message when the new poll slot appears in the polling cycle. On receiving JOIN message at CCU, it will create a new polling slot and data slot for the new node. In the same way when a node wants to leave the network it will do so by sending a DEPART

Central Controlling Unit
[CCU]
POLL
DATA frame
POLL+ACK
NACK
DATA frame
POLL+ACK
node1
node2
node3

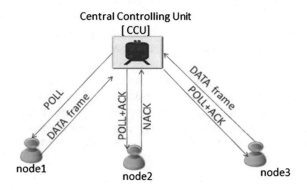

Fig. 4 A polling MAC based transmission sequence

Poll
DATA frame
Poll+ACK
NACK
....
DATA frame
Poll(join) +ACK
Join/Depart message
Cycle time

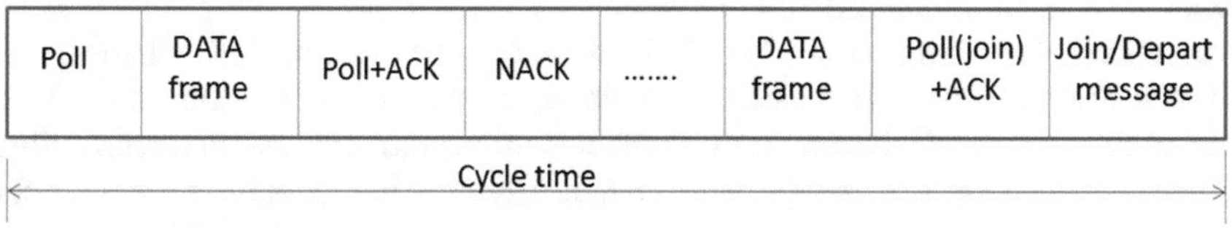

Fig. 5 Poll cycle for poll based transmission sequence

message to the CCU and the allocated poll and data slots will be removed from the polling cycle and hence the scalability is met. When a new node joins the BAN the polling cycle length is increased thus increasing the packet delay by a fixed value.

The poll based MAC protocol suffers from three main issues. The primary issue is that the number of control messages are large as it uses ACK/NACK and poll messages in each cycle that are equal to the number of nodes in the network. The next issue is in finding a suitable node that can act as the CCU, since CCU needs to have fast processing and large memory. Despite of the power constraint a node is to be determined that can take care of all the nodes in the network. Third and most important issue is that there is no provision of handling emergency data. The node having the emergency data will have to wait for its turn till CCU come and poll it for the data.

5.3 Random Access MAC Protocol

Random access MAC protocols are demand access protocols that are designed for short range communications. These protocols are best for the case with no or minimal control signalling. Whenever a node has some data to transmit it will check the status of the transmission channel, if it finds transmission channel is free then the node initiate a packet transmission using a contention based procedure. Main advantage of random access protocol in BAN is lower signalling overhead, scalability and no centralized control that allows the network to accommodate different data rates and inter-arrival rates. Priority can be assigned in this case. Consider a node generates some urgent data in a non-scheduled manner, then a TDMA or a polling based system will have difficulties to transmit that data immediately. But, in case of random access MAC protocol, data can be transmitted immediately provided the nodes have the necessary priority [23]. Drawback of this technique is that it can cause long delay if the traffic load is high, but in BAN most of the time the network traffic is stable.

$$T_{d(random)} = T_{access} + T_{retry} + T_{queue} + T_{packet} \qquad (1)$$

Equation 1 gives the average packet delay of a random access protocol based network. T_{access} represents the time required by a node to access the channel. T_{retry} represents the retransmission delay, if a packet experiences a collision then the transmitting node will backoff for certain duration and then try to transmit the packet. The backoff period will be randomly distributed to minimize the collision probability. There will be multiple retransmission attempts to resolve the collision, a packet could be dropped if the packet cannot be transmitted successfully after multiple collisions. T_{queue} represents the buffering delay at a node. T_{packet} value depends on the packets size and the transmission data rate. This protocol requires that nodes perform clear channel assessment (CCA) before transmission. But the CCA is not always guaranteed in BAN since the path lose inside the human body due to tissue heating is much higher than in free space. BAN nodes are energy constrained and the nodes have to keep track of the channel to get the turn of the transmission and hence it is not an idle MAC for BAN.

5.4 Preemption Enhanced Distributed Channel Access Priority MAC Protocol

In the context of time critical sensor applications, a protocol is required that gives emergency traffic privilege to interrupt other routine traffic in the network. Balakrishnan et al. [24] has proposed enhanced distributed channel access (EDCA) protocol that is based on priority and pre-emption concept. This scheme gives lowest delay and guarantees delivery of emergency frames, even when the traffic load is high. However, this causes increase in delay of normal traffic that is allowed in case of emergency. Parameters used in the protocol are short inter frame space (SIFS), emergency priority SIFS (EPSIFS), emergency priority arbitrary inter frame space (EPAIFS), and normal priority SIFS (NPSIFS). EPSIFS is designed to meet the emergency. In case of emergency all frames are separated by EPSIFS, so that emergency flow of data can never be interrupted [23]. EPAIFS is a combination of EPSIFS and CCA whereas NPSIFS is a combination of EPAIFS and CCA). These have values that are related to each other as $SIFS < EPSIFS < EPAIFS < NPSIFS$.

The basic concept behind this is similar to the IFS design of 802.11 schemes, that uses the wait times to allow for precedence in acquiring the free channel. As shown every transmission cycle consists of (Data-Ack-Data) is separated by different IFS as shown in Fig. 6 that give them the precedence irrespective of the frame priority [25].

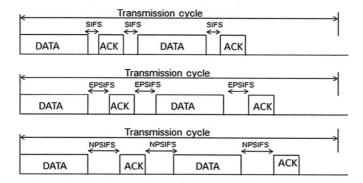

Fig. 6 Inter frame spacing used in preemption-EDCA MAC

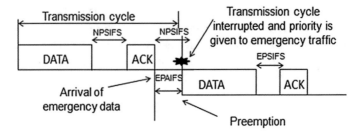

Fig. 7 Preemption in EDCA MAC

Figure 7 gives detailed explanation of P-EDCA working, guaranteeing medium access for emergency priority through transmission preemptions. Emergency frame sequence is separated by the shortest wait time, new emergency frames start channel acquisition or contention if they sense the channel free for an EPAIFS period, which is smaller than NPSIFS. The Data-Ack sequence is never interrupted and can be achieved by setting the network allocation vector (NAV) to the end of ack transmission.

Although this preemption based protocol has solution of priority transmission problem, but it has the obvious shortcoming. The primary shortcoming is due to frequent transmission of emergency data. If transfer of emergency data is frequent and take a large time, then the other nodes which want to transfer regular data have to wait. This results in waste of their memory and energy. So, the working time of the nodes will be high and it will decrease the lifetime of the full system. Secondly, this scheme cannot solve the hidden node problem in BAN. Preemptions by hidden nodes will be getting bad result for emergency information transmission due to collisions. Additionally, because of channel errors if there is a requirement of re-transmissions because of preemptions it will delay performance by this scheme.

5.5 Sensor MAC

Sensor MAC (S-MAC) is a typical synchronous MAC protocol in which nodes send their schedule to the neighbour nodes, when they will be available for communication [26]. S-MAC has a periodic listen and sleep cycle. In sleep mode nodes turn off its radio and in wakeup mode node is ready to exchange its information. This sleep cycle duration stays constant during the lifetime of the network and is determined by the duty cycle. Transmission frame of S-MAC can be divided into two parts such as DATA and SYN. In SYN part, the sensor nodes update their neighbour tables to determine which node will communicate at what time and in DATA part it actually perform the data communication.

The disadvantage of S-MAC is that the value of duty cycle is fixed for the lifetime of the network and if a new sensor node is to be added in the network, full network communication structure needed to be re-established. In addition, S-MAC can efficiently handle periodic traffic but there is no provision to handle aperiodic or emergency traffic [27].

5.6 Traffic Adaptive MAC

Contention based protocol for BAN will perform badly in most of the cases as most of the traffic is correlated. For example, a patient suffering from fever triggers temperature, blood pressure, respiration sensors at the same time. Additionally, contention based protocol cause heavy collision and extra energy consumption. Traffic adaptive MAC protocol (TaMAC) for BAN is an energy efficient protocol that uses traffic information of the nodes. In BAN, traffic can be classified as: normal, emergency and on-demand traffic. TaMAC uses a star topology where a central coordinator controls the entire operation of the network. It is the duty of the coordinator to schedule flow of traffic using traffic patterns. This traffic pattern is used only for normal traffic. In case of emergency or on demand traffic, radio wakeup mechanism is used [28].

As shown in Fig. 8, frame contains configurable contention access period (CCAP) that consist of a few minislots of equal duration for short data transmission and CSMA/CA or slotted ALOHA is used for such transmissions. A contention free period (CFP) that consist of guaranteed time slots and are used for actual data transmission. This duration is configurable according to the traffic pattern. The coordinator maintains a table with identity of each nodes and their corresponding traffic pattern. To handle periodic data, each node has a predefined time slot, they wake up only when they have to do some communication. For emergency traffic, the node which wants to send emergency data will send a wakeup radio signal to the coordinator. For ondemand traffic the coordinator send a wakeup radio signal followed by a beacon frame, this mechanism is also used for updation of traffic wakeup table.

The advantage of TaMAC is that it can efficiently handle emergency traffic. In addition, wakeup radio signal is used to wake up the main data channel for data transmission. Radio channel is mainly used for the wakeup of data channel and its power requirement is limited.

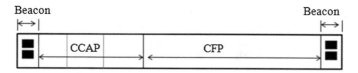

Fig. 8 Frame structure for traffic adaptive MAC

5.7 X-MAC

Protocols that use low power listening (LPL) and uses asynchronous data cycle, must use preamble for managing transmission cycle. These introduced protocols have several disadvantages. Primarily the receiver has to wait for the full period of preamble for DATA/ACK exchange. Even if the receiver wakes up at the start of preamble, it wastes energy both at sender and receiver ends. Secondly, preamble based MAC technique also suffers from overhearing problem. It is because, all receivers are getting preamble wake up and waits for preamble to end, to find out whether they are the targeted node or not. It waste a lots of energy of a group of exposed nodes. The third disadvantage is that the overall lifetime of the network is degraded. It is because more than one nodes are involved in relay based routing [23]. This phenomenon is depicted in Fig. 9.

X-MAC tries to eliminate these shortcomings by making use of a short preamble. To improve the energy efficiency, address of the targeted node is embedded in every preamble so that other nodes can go to sleep quickly. Length of preamble cycle should be maintained greater than the maximum length of the receiver sleep cycle. If a node want to send data but detects a preamble, it can determine the address of the intended node (say node A). If some node is also interested in communicating with the node A, after hearing the ACK from node A, and following a random back-off

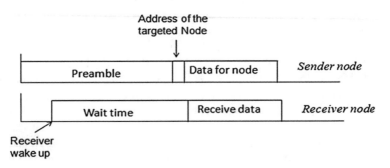

Fig. 9 Long preamble causes energy wastage

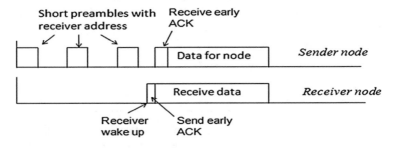

Fig. 10 Use of short preamble and early ACK in X-MAC

it can directly start sending data to the node without any preamble. The use of short preamble X-MAC protocol is shown in Fig. 10 and is mainly designed for efficiently handling of asynchronous data.

5.8 Medical MAC (M-MAC)

Some nodes in BAN are more energy conservative like in-body nodes as compared to those which are present on the body. So more functionality can be assigned to outside nodes and hence medical MAC (M-MAC) uses the concept of master-slave. In M-MAC, body nodes periodically send sensors reading to a central outside master node [8, 29]. There will be a number of master slave clusters and possibility of collision is avoided between various clusters using the clear channel assessment (CCA). A master slave relation between nodes in M-MAC is shown in Fig. 11. Three type of communication can take place in M-MAC. These are listed below.

1. Master node keeps on sending beacon signal periodically so that new nodes in its range can respond immediately. Further, slave joins the cluster of the master with strongest power of the received signal.
2. Both master and slave manage their sleep cycle so that they wake up at the prescheduled time. Master asks slave for the sensed data and gives it next wakeup slot. As the number of slaves in the cluster changes occasionally, simple TDMA is used within a cluster.
3. In case of emergency, slave communicates directly with the master without waiting for the starting of next preassigned slot.

There are many other MAC protocols which perform good for some parameters but fail for others. Tables 3 and 4 gives a brief comparison of already discussed protocols on various parameters.

Fig. 11 Master slave nodes in medical MAC

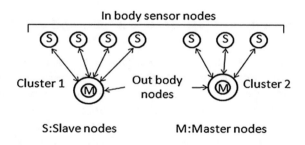

In body sensor nodes

Cluster 1   Out body nodes   Cluster 2

S:Slave nodes   M:Master nodes

Table 3 Comparison of various MAC protocols for BAN on various features

Features	Scheduled TDMA MAC protocol	Polling MAC protocol	Random access MAC protocol	Preemption-EDCA MAC
Energy consumption	High in case of non periodic traffic	High but better than TDMA	Much better energy efficiency, node wake up only when it has some data to transmit	High because of preemption retransmission may be required often
Transmission overhead	Very low or no overhead	High overhead due to polling	Overhead is high unless aggregation approach is applied	High overhead
Network scalability	Very poor, worst for non-periodic traffic	Scalability is good and is limited by the delay	Good	Best
Traffic handling capability	Efficiently handle periodic traffic, but fails for non periodic traffic	Efficiently handle periodic traffic by adjusting polling cycle, can cope up with aperiodic traffic	Both periodic and non periodic traffic can be handled in a good way	Best for both periodic and non periodic traffic
Packet loss	Deterministic and fixed	Deterministic and vary with load in network	Variable and vary with type of conditions of patient	Variable and vary with type of conditions of patient
Memory consumption	Low	Low	Moderate	High
Acknowledgment	No	No	No	Yes

6 Implications for Researchers and Practitioners

This chapter has discussed various MAC protocols for BAN with its advantages and disadvantages. As MAC layer plays a vital role for improving the lifetime of any BAN. Designing an efficient MAC protocol is the open challenge for practitioners and researchers. Fixed time slots MAC protocol can be modified to manage emergency data, so that advantages of TDMA can be incorporated in BAN. Work can be done to reduce the polling message in poll based MAC. The sleep cycle of the nodes having no data to send should be managed effectively. A technique needs to be devised for minimizing the delay of packet delivery in each type of MAC protocols especially in random access protocol. Many challenges related to retransmission are also open like rescheduling keeping energy conservation in mind. Finding relay nodes by not affecting overall lifetime of the BAN [30]. Finding proper transmission power that

Table 4 Comparison of various MAC protocols for BAN on various features

Features	S-MAC	Ta-MAC	X-MAC	M-MAC
Energy consumption	Energy efficient	Highly energy efficient	Highly energy efficient	Energy efficient for in-body nodes but not for on body nodes
Transmission overhead	No polling overhead so low overhead	Very low	Intermediated due to use of preamble/short preamble	Intermediated as the nodes are divided into clusters
Network scalability	Very poor	Best	Good	Good
Traffic handling capability	Efficiently handle periodic traffic, but fails for non periodic traffic	Best for both periodic and non periodic traffic	Best for non-periodic	Efficient for periodic, but not as good for emergency traffic
Packet loss	Deterministic and fixed	Variable and vary with type of conditions of patient	Deterministic and fixed	Variable and vary with type of conditions of patient
Memory consumption	High	High	Moderate	Low
Acknowledgment	No	Yes	Yes	No

is in safe range for human and handling of both periodic as well as emergency traffic are to be sorted out in developing any new MAC protocol for BAN. Although IEEE 802.15.6 is present, but it needs to be modified according to the need and specific requirements of BAN.

7 Conclusion

BAN is an emerging area of networking that can help human in many ways varying from healthcare to entertainment. Still it is in initial stage and once it gets matured it will give breakthrough invention in medical, military training, vascular network, consumer electronics etc. The nodes associated in the network are extremely resource constrained in terms of storage and energy, also have unique wireless channel that have different set of challenges to be sorted out. The strength of a BAN signal is affected by the physical location (in or on body) and orientation of the nodes, in relation to each other as well as the human body. Simple protocols like TDMA, S-MAC has low overhead but have very low scalability, TaMAC, which is good in network scalability has large memory consumption, M-MAC which require less memory and uses master-slave concept in energy efficient for in-body nodes but not for on-body nodes. Therefore, there is a need to find a protocol which is energy efficient for both in and on body sensors, have low transmission overhead, does not degrade with increase in number of nodes, and have low and deterministic packet loss.

References

1. Chen, M., Gonzalez, S., Vasilakos, A., Cao, H., Leung, V.C.M.: Body area networks: a saurvey. J. Mob. Netw. Appl. **16**(2), 171–193 (2011)
2. Ullah, S., Higgins, H., Braem, B., Latre, B., Blondia, C., Moerman, I., Saleem, S., Rahman, Z., Kwak, K.S.: A comprehensive survey of wireless body area networks: on phy, mac, and network layers solutions. J. Med. Syst. **36**(3), 1065–1094 (2012)
3. Jovanov, E., Milenkovic, A., Otto, C., Groen, D.: A wireless body area network of intelligent motion sensors for computer assisted physical rehabilitation. J. Neuroeng. Rehabil. **2**(6), 1–10 (2005)
4. Khan, J.Y., Yuce, M.R.: Wireless body area network for medical applications. In: Domenico Campolo (ed) New Developments in Biomedical Engineering, pp. 591–627 (2010). http://www.intechopen.com/books/new-developments-in-biomedical-engineering/wireless-body-area-network-wban-for-medical-applications
5. Quwaider, M., Biswas, S.: Physical context detection using wearable wireless sensor networks. J. Commun. Softw. Syst. **4**, 191–201 (2008)
6. Polastre, J., Hill, J., Culler, D.: Versatile low power media access for wireless sensor networks. In: Proceedings of 2nd ACM Conference on Embedded Networked Sensor Systems, p. 95107 (2004)
7. Ullah, S., Higgins, H., Islam, S.R., Khan, P., Kwak, K.S.: On PHY and MAC performance in body sensor networks. EURASIP J. Wirel. Commun. Netw. **1**, 1–7 (2009)
8. Otto, C.A., Jovanov, E., Milenkovic, A.: A WBAN-based system for health monitoring at home. In: Proceedings of 3rd IEEE/EMBS International Summer School on Medical Devices and Biosensors, pp. 20–23 (2006)
9. Dong, J., Smith, D.B.: Cooperative receive diversity for coded GFSK body-area communications. Electron. Lett. **47**(19), 1098–1100 (2011)
10. Schwiebrt, L., Gupta, S.K., Weinmann, J.: Research challenges in wireless networks of biomedical sensors. In: Proceedings of 7th Annual International Conference on Mobile Computing and Networking, pp. 151–165 (2001)
11. Boulis, A., Smith, D., Miniutti, D., Libman, L., Tselishchev, Y.: Challenges in body area networks for healthcare: the MAC. IEEE Commun. Mag. **50**(5), 100–106 (2012)
12. Zhen, B., Li, H.B., Kohno, R.: IEEE body area networks and medical implant communications. In: Proceedings of the ICST 3rd International Conference on body Area Networks, p. 26 (2008)
13. Penders, J., van de Molengraft, J., Brown, L., Grundlehner, B., Gyselinckx, B., Van Hoof, C.: Potential and challenges of body area networks for personal health. In: Proceedings of Annual IEEE International Conference of the Engineering in Medicine and Biology Society, pp. 6569–6572 (2009)
14. Bugler, G.: Communication protocols for multi-hoping wireless body sensor network. B. Tech. thesis, University of Newcastle, Australia, pp. 1–20 (2008)
15. Lorincz, K., Malan, D.J., Fulford-Jones, T.R.F., Nawoj, A., Clavel, A., Shnayder, V., Mainland, G., Welsh, M., Moulton, S.: Sensor networks for emergency response: challenges and opportunities. IEEE Pervasive Comput. **3**(4), 16–23 (2004)
16. Tselishchev, Y., Libman, L., Boulis, A.: Energy-efficient retransmission strategies under variable TDMA scheduling in body area networks. In: Proceedings of 36th IEEE Conference on Local Computer Networks, pp. 374–381 (2011)

17. Penders, J., Van de Molengraft, J., Brown, L., Grundlehner, B., Gyselinckx, B., Van Hoof, C.: Potential and challenges of body area networks for personal health. In: Proceedings of Annual International Conference of the IEEE Engineering in Medicine and Biology Society, pp. 6569–6572 (2009)
18. Li, H.B., Takizawa, K., Kohno, R.: Trends and standardization of body area network for medical healthcare. In: Proceedings of European Conference on Wireless Technology, pp. 1–4 (2008)
19. Yazdandoost, K.Y., Sayrafian-Pour, K.: Channel model for body area network. IEEE P802.**15**, 08–0780 (2009)
20. Fang, G., Dutkiewicz, E.: BodyMAC: energy efficient TDMA-based mac protocol for wireless body area networks. In: Proceedings of 9th IEEE International Symposium on Communications and Information Technology, pp. 1455–1459 (2009)
21. Ullah, S., Khan, P., Choi, Y.W., Lee, H.S., Kwak, K.S.: MAC hurdles in body sensor networks. In: Proceedings of 11th International Conference on Advanced Communication Technology, vol. 2, pp. 1151–1155 (2009)
22. Ye, W., Silva, F., Heidemann, J.: Ultra-low duty cycle MAC with scheduled channel polling. In: Proceedings of the 4th International Conference on Embedded Networked Sensor Systems, p. 321334 (2006)
23. Buettner, M., Yee, G.V., Anderson, E., Han, R.: X-mac: a short preamble mac protocol for duty-cycled wireless sensor networks. In: Proceedings of the Fourth International Conference on Embedded Networked Sensor Systems, p. 307320 (2006)
24. Balakrishnan, M., Benhaddou, D., Yuan, X., Gurkan, D.: Service preemptions for guaranteed emergency medium access in wireless sensor networks. In: Proceedings of IEEE Military Communications Conference, pp. 1–7 (2008)
25. Kim, T., Lee, H., Koh, J., Lhee, K.S.: A performance analysis of polling schemes for IEEE 802.11 MAC over the GilbertElliot channel. Int. J. Electron. Commun. **63**(4), 321–325 (2009)
26. Ye, W., Heidemann, J., Estrin, D.: An energy-efficient mac protocol for wireless sensor networks. In: Proceedings of 21st Annual Joint Conference of the IEEE Computer and Communications Societies, vol. 3, pp. 1567–1576 (2002)
27. Nabi, M., Basten, T., Geilen, M., Blagojevic, M., Hendriks, T.: A robust protocol stack for multi-hop wireless body area networks with transmit power adaptation. In: Proceedings of Fifth ACM International Conference on Body Area Networks, pp. 77–83 (2010)
28. Laneman, J.N., Tse, D.N., Wornell, G.W.: Cooperative diversity in wireless networks: efficient protocols and outage behaviour. IEEE Trans. Inf. Theory **50**(12), 30623080 (2004)
29. Omeni, O., Wong, A.C.W., Burdett, A.J., Toumazou, C.: Energy efficient medium access protocol for wireless medical body area sensor networks. IEEE Trans. Biomed. Circuits Syst. **2**(4), 251–259 (2008)
30. Cao, H., Leung, V., Chow, C., Chan, H.: Enabling technologies for wireless body area networks: a survey and outlook. IEEE Commun. Mag. **47**(12), 8493 (2009)

Internet of Nano Things and Industrial Internet of Things

Hemdan Ezz El-Din and D.H. Manjaiah

Abstract In recent years, nanotechnology has become an important research topic which promises novel solutions for several applications in healthcare, industrial, biomedical, and military. The advanced nanotechnology led to appear new nanodevices which acquire, generate, compute, process and transfer data at nanoscale dimension. These nanodevices are interconnected with each other using the existing communication systems which produce a new domain that is called Internet of nano things. The current advanced development in telecommunication and network system let to new area known as Internet of things. One of the most important applications of Internet of things is in the industrial control systems. The Internet of things makes novel development in industrial field which define new paradigm that is further referred to as industrial Internet of things. Within this context, this chapter through light on fundamental concepts, architecture, communication classifications, communication issues, applications, security and future research directions in the Internet of nano things. Additionally, it also explore architecture, requirements, benefits, security and future research directions in the industrial Internet of things.

1 Introduction

Internet of things (IoT) has become a vital research topic for researchers, organizations and companies by providing large investments in this area as an important future direction in network and communication systems [1–3]. This will enable things, machines and people to communicate and interact to each other in efficient and effective way. The IoT defines a new paradigm to interact and communicate of the real world physical devices such as sensors, actuator, home appliances, mobiles and many other elements in seamless way.

H. Ezz El-Din (✉) · D.H. Manjaiah
Department of Computer Science, Mangalore University, Mangalore, India
e-mail: ezzvip@yahoo.com

D.H. Manjaiah
e-mail: manju@mangaloreuniversity.ac.in

© Springer International Publishing AG 2017
D.P. Acharjya and M. Kalaiselvi Geetha (eds.), *Internet of Things:
Novel Advances and Envisioned Applications*, Studies in Big Data 25,
DOI 10.1007/978-3-319-53472-5_5

Recent time is the era of nanotechnology which introduce the concept of nanodevices or nanomachines which are in a scale ranging from one to nanometers in dimension. Also, the nanotechnology led to new nanomaterials which have new properties and characteristics that will make novel developments in nanodevices like nanosensor and nanorouter. These nanodevices are integrated to perform different tasks such as sensing, acquisition or transferring data through nanonetwork. The nanonetwork will cover unmatched locations to perform additional in network processing. The emergency development technology that develop and create nanomachines defines a new networking paradigm that consists of interconnection of nanoscale devices with existing communication networks and ultimately the Internet which is called Internet of nano things (IoNT). The IoNT add new dimension for IoT technology by embedding nanosensors inside the devices enabling them to communicate through the nanonewtwork via the Internet for global connection among a lot of devices around the world.

One of the emerging trends in embedded systems industry is the Internet of things which have much important innovation in industrial field that led to new domain which is called industrial Internet of things (IIoT). The IIoT is a network paradigm consists of physical elements, platforms and software to communicate and share data between them in smart manner. The corporation between the IIoT and existing industrial technologies will improve and facilitate capture of and access to real-time information.

The reminder of this chapter is organized as follows. Section 2 introduces architecture, communication classifications, communication issues, applications, security and future research directions in the Internet of nano things. Overview, architecture, requirements, benefits, security and future research directions in industrial Internet of things are discussed in Sect. 3. Chapter conclusion is presented in Sect. 4.

2   Internet of Nano Things

Nanotechnology led to new nanomaterials with new properties and characteristics that will make novel developments in nanodevices such as nanosensor and nanorouter. IoNT is the interconnection of nano-scale devices with the exsiting telecommunication system and networks [4]. The IoNT is added new dimension for the Internet of things by embedding nanosensors inside the devices which enable them to communicate together through the nanonewtwork via the Internet for global connection among devices around the world. The IoNT describes how the Internet will get bigger as nanosensors which are connected to physical things such as physical assets or consumer devices for collecting, processing and sharing of data with the end-users. IoNT has many vital applications such as smart agriculture, healthcare, military, logistics, aerospace, industrial control systems and manufacturing and smart cities. There are new domains promised from the IoNT technology such as Internet of Bio-Nano Things (IoBNT) [5] and Internet of Multimedia-Nano Things (IoMNT) [6] which will make new developments in healthcare and multimedia fields.

2.1　Architecture of Internet of Nano Things

The Internet of nano things consists of underlying nanonetworks which are connecting nanosensors and nanodevices to communicate and interact with each other in distributed manner. The interconnection of nanodevices in the IoNT requires developing of new networking components to perform the communication process in nanonetworks. The architectures of IoNT consist of the following components as discussed below [4, 7].

1. *Nano-end points*: These points are nano-nodes such as nanosensor and nanoactuator devices which are able to perform simple processing and computing process. They have limited memory size that makes them to transmit only over very short distances due to their reduced energy and limited communication capabilities. These devices can be included inside human body in addition to ability to integrate in many types of things such as books and keys which can be considered as nano-end points.
2. *Nano-routers*: Nano-routers have comparatively larger processing resources than nano-nodes and they are suitable for collecting and acquiring information coming from nanosensor. Nano-routers can also control the behaviour of nano-end points by exchanging very simple control commands.
3. *Nano-micro interface devices*: Nano-micro interface devices are able to collect and aggregate information and data coming from nano-routers, to convey it into the microscale, and vice versa. Nano-micro interfaces may be as hybrid devices able to communicate in the nanoscale using nanocommunication techniques and to use classical communication paradigms in conventional communication networks.
4. *Micro-gateway*: Micro-gateway will use to enable more remote control of the nano-devices over the network (i.e. Internet). This micro-gateway can receive and forward data inside the nanonetwork.

For new domains in the IoNT such as multimedia nano things and bio-nano things, the architecture of nanodevices will be different that are depend on nanotechnology. For example, in internet of multimedia nano things (IoMNT), the multimedia nanodevices consist of the following componets; nano-camera, nano-phones, nano-antenna, nano-processors, nano-power unit and nano-EM transceiver as depicted in Fig. 1.

2.2　Communication Classifications

Communication is the process of transfering data and information between different components inside a network. In the Internet of nano things, the situation will be different because the use of nano scale devices that led to the appearance of nanonetworks. Nanonetworks are connecting nanodevices which can perform sensing, collecting, processing and storing information. Classical communication

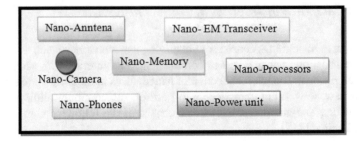

Fig. 1 Multimedia nano-device elements

paradigms need to be revised for being suitable for nanodevices (i.e., nanomachines) to communicate in the nanoscale networks. There are two types of communication in the nanoscale networks as stated below [4].

Molecular Communication (MC): This type of communication is defined as the exchange of information through the transmission and reception of molecules. These molecules will interact with nanodevices in biological environment such human body.

Electromagnetic Communication (EM): This type of communication is defined as the exchange of information through the transmission and reception of electromagnetic radiation from nanodevices in nanonetworks. These radiations will be emitted in specific bandwidth for allow nanodevices to interact and communicate to each other.

2.3 Communication Issues

This section provides communication issues in the Internet of nano-things to facilitate communication process between nanodevices in nano-networks. There are many communication issues that can be taken into consideration such as nanodevices addressing, information routing, sharing of communication channel, network discovery, reliability issues [6, 8].

Nanodevices Addressing: Assigning an address to every nanodevice in nanonetwork require synchronization between nanodevices. Each nanodevice which works as end-point in the nanonetwork should have unique address to prevent collision during communication between these nanodevices. Nanonetwork contains large number of nanodevices that cannot be easily managed without giving each nanodevice a unique address. An example of addressing nanodevice in nanonetworks is: an address such (G7.I8.R2.N5), can be used to refer to the nanonode 5, within the domain of the nano-router2, connected to nano-micro interface 8, linked to gateway 7.

Information Routing: Establishment routes to allow data and information to be exchanged among nanodevices in the nanonetwork. It requires new standards and protocols to manage the transmission and reception of information in easy manner without any complication. With increasing number of nano-nodes inside the nanonetwork the communication process will be more complex to route information between large numbers of nano-nodes. Nano-routers will be used for routing information in the nanonetworks and for aggregating information which come from limited nanosensor devices in addition to control the behaviour of nano-end points by exchanging very simple control commands.

Sharing of Communication Channel: Mechanisms for sharing communication channels to access them which are needed to achieve good communication in nanonetworks. There are many mechanisms which cannot be used in pulse based communications such as carrier sense medium access control (CSMA) because there is no carrier signal to sense. The CSMA is a probabilistic media access control protocol in which a node verifies the absence of other traffic before transmitting on a shared transmission medium, such as an electrical bus, or a band of the electromagnetic spectrum. Synchronization during communication between nanonetworks is also required.

Reliability Issues: Reliability is an important issue to guarantee the transfer message from remote control center to nanodevice in the nanonetwork. The reliability of network can be affected by many aspects such as nanodevice failure and transient molecular interference in the channel and unexpected errors from nanodevices during the process of exchanging data and information.

Network Discovery: In nanonetworks every nanodevice is expected to connect to all other nanodevices in the nanonetwork and discover all surrounding devices to interact and communicate with them so that there is a need to new solutions for network discovery to discover any new nanodevice when they are connected to the nanonetwork.

2.4 Applications

Internet of nano things has many applications in several fields such as multimedia, military, biomedical, industrial and environmental science as shown in Fig. 2. A brief discussion on these applications is presented below. Details on these applications can be found here [5, 6, 8].

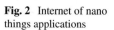 **Fig. 2** Internet of nano things applications

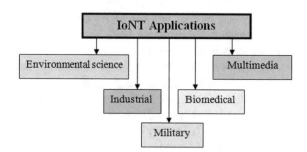

Multimedia: Nanotechnology proposed novel nonmaterial which can be developed to use the manufacturing of a new generation of miniature photodetectors and acoustic nano-transducers, which can generate multimedia content at the nanoscale devices for multimedia applications such as ultra-high-resolution imaging for crime scenes and ultra-high-resolution imaging of distant objects for satellite imaging. To capture visual and acoustic information with unprecedented resolution and accuracy, much higher than current micro-cameras and microphones, so that there is a serious need to propose new types of nano-cameras and nano-phones. These nano-devices will overcome the shortages of current multimedia devices by providing higher quality image and audio sensing capabilities, higher computational and storing capacities and higher energy efficiency.

Military: The war not like before, now advanced biological and chemical weapons that have huge destroy capabilities which can be used. In this domain, the IoNT can use nanosensors that have the ability to detect the presence of a chemical composite in a concentration as low as one molecule. These nanosensors can recover the molecular composition of a room or the battlefield without requiring external and large equipment such as the devices used for spectroscopy. These nanosensors that connected together can be used to detect the origin of problems of very small cracks in bridges, civil structures, vehicles, textiles and rockets.

Biomedical: Nanosensors now are used for Bio-Nano field for delivering a drug in specific devices of human body when it has any problem and monitor the level of a specific substance. In addition to control the intra-cranial pressure, a chemical compound to dissolve a clot in an artery. Embedding nanosensors in human body enable doctors and healthcare provider to remotely control and monitor any unexpected change in the body devices by collecting information through these nanosensors. Nanosensors also enable healthcare provider and doctors for monitoring the glucose level in blood and different infectious agent for dangerous diseases like cancer by checking biomarkers of cancer.

Industrial: Internet of nano things can be applied for many domains in industrial control systems such as enhancing sensitivity of touch technology from surfaces by using air through developing novel nanosensors that can detect figure movements in the air and convert it into signals and this technology known as air touch technology (ATT) that will change the touch technology industry in the next years for all electronic devices and also use eyes glass instead of using cell phones which have large size. Nanotechnology will help to develop these new trends in touch technology. Also, nanodevices will use for implementing future interconnected office with the existing communication technologies to improve the level of interacting people together in work environment.

Environmental Science: The high sensibility and large diversity of chemical nanosensors can be exploited in several environmental applications that are otherwise unfeasible with current technologies. In environmental science nanodevices such as nanoactuators and nanosensors can be used in monitoring and controlling natural processes between plants. There are many types of biological and chemical nanosensors that can be used to detect the chemical compounds that are being released and exchanged between plants. The interaction of nanodevices with natural plants can be remotely done from environmental and agricultural centre of studies for research purpose and develop new products and enhance the existing output of surrounding environment.

2.5 Security Issues in IoNT

In the Internet of nano things, security plays a vital role in providing safe and reliable communication environment between nanodevices in nanonetworks. Nanonewtorks consist of nanodevices that can connect together for exchanging of information. Attackers can exploit vulnerabilities and weakness in the nanonetwork. Current security mechanisms and techniques cannot be suitable to secure nanodevices in the nanonetwork from malicious attacks and crimes because nanodevices are working in terahertz band physical layer. To protect the IoNT infrastructure, there is a serious need to propose and develop new security solutions to prevent crimes related to the IoNT. The existing security solutions cannot be used directly for securing the IoNT infrastructure. Some of the suggested solutions for securing the IoNT environment are; check the integrity of data by using checksum algorithms, use encryption algorithms to encrypt data before transfering between nanodevices, use data hiding algorithms for hide critical data and use multi-layer authentication to guarantee authenticate user only can access nanonetworks.

2.6 Future Research Directions in IoNT

The advances in nanotechnologies will open new research direction to enable the internet of nano things in our daily life. There are many research points that can be taken into consideration as below.

1. *To design and develop nanorobots in microscale dimension*: Nanotechnology has become one of the most important technologies to develop nano-scale devices for many fields like medical and healthcare. These nanodevices can be integrated together to produce nanorobotics. The nanorobotics is a field to develop and create robots which are in nano-scale dimension.
2. *To develop new transmissions mechanisms*: In the nanonetwork, the nanodevices can communicate in distance in range less than one meter which requires new modulation mechanisms to handle communication of information between the nanodevices in efficient and effective manner.
3. *To develop new compression techniques*: The compression process is important for reducing size of data before transmission in nanoectwork to save power and energy consumption in addition to transfer more amount of data in short time.

4. *To design new architecture for nanodevices*: Nanodevices are the main element of the IoNT infrastructure these devices can be useful for many fields like biological and multimedia.
5. *To develop new security mechanisms and techniques*: Security is an important issue to provide secure communication channels to transfer data and information between nano-scale devices inside the nanonetworks.
6. *To develop algorithms for power consumption*: Power and energy consumption help to make nanodevice to life more time especially in critical application like nanodevice in human body.
7. *To develop very fast communication mechanisms*: It is necessary to develop these types of mechanisms to be suitable nano-things.
8. *To develop mechanisms for coding of communication channels*: Developing mechanisms to be able to adapt with transmission of information in terahertz band in addition to reduce channel errors are necessary in nanonetworks.
9. *To develop new types of antennas*: In the nano-communication, it is important to design and develop special antennas to support transmission and reception of information between nanodevices.
10. *To develop routing protocols*: Routing protocols are responsible for making transmission of data between sender and receiver easier by chossing the best route between them. Developing new routing protocols will be an effective solution in nano-scale environment (i.e.IoNT).
11. *To develop new digital forensic techniques*: The digital forensic is the process of extract digital evidence from digital devices. In nanonetwork, developing new techniques and methodology will help to discover weakness in addition to trace attackers.
12. *To develop analytical techniques*: Analyze data which are generated from nanodevices will help to predict and make decisions related to the application in which the data are generated.
13. *To develop privacy techniques*: These techniques will help to protect and secure sensitive and critical information.

3   Industrial Internet of Things

In industrial space, there are many limitations to make industrial environment smarter such as reduce power consumption and efficient operational work inside industrial factories. Now, overcoming these limitations became possible thanks to new technologies like internet of things and nano-things. The IoNT is an interconnection of nano-scale devices with accessible Internet and communication networks. It is embedded with nanotechnology which help in seamless transmission and communication of data within a given range of operations that can be applied for industrial field that create new domain called industrial Internet of things (IIoT). The IIoT is a network paradigm which consist of physical elements, platforms and software to communicate and share data between them in intelligence way. The corporation

Fig. 3 Relation between
internet of things and
industrial internet of things

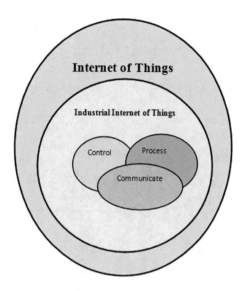

between IIoT and existing industrial technologies will improve and facilitate, capture of and access to real-time information. In addition to this, it will help companies
and organizations to develop their business and open up a new era of economic
growth and competitiveness. Using intelligent technologies in the IIoT will have a
great impact on increasing productivity, efficiency and operations of industries in the
world. There are many applications available in our life such as a wind farm which is
equipped with sensors and other hardware components to communicate and interact
with all the other windmills.

In the recent years, IIoT and IoT are two important technologies which are helping
to improve and enhance the level of people life. There is a relation between IIoT
and IoT technologies as shown in Fig. 3. The Internet of things has an imperative
economic and societal impact for the future construction of communication and
network system to exchange information between people and things.

3.1 Functional Model of Industrial Internet of Things

Industrial internet of things contains four parts: intelligent assets; a data communication infrastructure; analytics and applications to interpret and act on the data, and
people. The Internet of things reference architecture provides a reference for building
compliant IoT architectures. As such, it provides views and perspectives on different architectural aspects of concern to IoT stakeholders. Within the IoT Reference
Architecture, a Functional view has been developed. It includes seven main areas of
functionality as shown in Fig. 4.

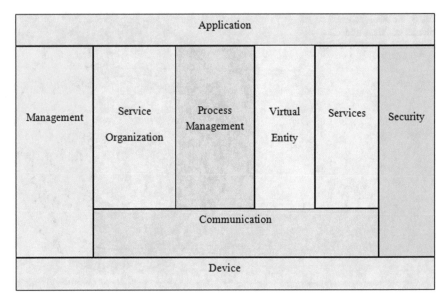

Fig. 4 Internet of things functional model

Communication: An abstraction, modeling the variety of interaction schemes derived from the many technologies belong to IoT systems and providing a common interface to the IoT Service. This component provides a reference stack for communicating with the intelligent devices.

Service: Services includes functionalities for discovery, look-up, and name resolution of IoT Services. This component provides for exposing device and sensor data as a service.

Virtual Entity: Functions for interacting with the IoT system on the bassis of virtual entities. This component provides for asset-based information exchange.

Process Management: Process modelling, process execution. This component provides an environment for modelling IoT-aware processes and the tools necessary to model business processes. It also executes these processes by utilizing IoT services orchestrated in the Service Organization Layer.

Service Organization: Service composition, orchestration and choreography. This function resolves the appropriate services that can handle the IoT Users request, and provides an asynchronous way to request service orchestration.

Security: Functions for ensuring the security and privacy of IoT compliant systems. This component provides for authorization, authentication, Identity management, key exchange and management for secure communications, and the like.

Management: This component provides functionalities for dealing with configuration, fault identification and isolation, performance, membership management, reporting, and state monitoring, prediction, and enforcement.

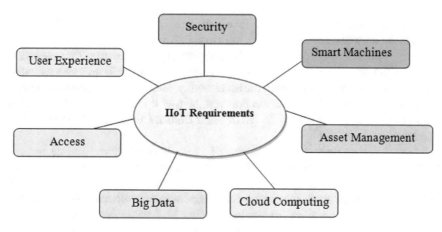

Fig. 5 Industrial internet of things requirements

3.2 Requirements for Industrial Internet of Things

Industrial Internet of things requirements are coming from the requirements of Internet of things. These requirements help to achieve the main goal of industrial IoT to improve industrial control systems by making industrial system intelligent and smarter in several applications like machines auto-diagnosis, temperature monitoring, manufacturing products, monitoring of ozone levels, etc. the Industrial IoT business needs cheap and inexpensive devices that work easily to install links like wired, wireless, power line, and network of the control system in addition to ability to work in safe and reliable manner. Some of the essential requirements in IIoT technology like big data and cloud computing are required for better performance and productivity. These requirements is depicted in Fig. 5.

Cloud Computing: Cloud computing is an attractive area for researchers working in industrial systems to enable the industrial IoT through using massive cloud capabilities such as computing, processing and storage. In the IIoT environment, devices like sensors can generate large amount of real time information that require huge processing and storage capabilities. These capabilities that can be provided by the cloud to achieve the cloud on demand, fast, efficient, and flexible data center to enable the IoT services.

Access: Access to IIoT infrastructure is an important to give users the ability to manage and control systems from anywhere and at anytime in timely fashion manner with take in consideration security issues.

Security: Security plays a vital role in IIoT for the safe and reliable operation of end to end connected devices. Secure wireless and wired communications in the industrial environment is very import to satisfy availability and safety parameters.

Big Data: There are large amount of generated data from devices and sensors in the industrial systems which need suitable techniques to manage and handle them like big data solution (i.e. Hadoop) to predict and enhance productivity and decision making. This will make more competitive in the industrial market for providing better service for organizations and users in effective and efficient way.

User Experience: User Experience is an important process to make the interaction between the user and product through enhancing user satisfaction. The user satisfaction can be done by improving the accessibility, and usability, entertainment and pleasure provided in the interaction between the user and the product. The process of user interaction includes the design of better Human-Computer Interaction (HCI) methods.

Assets Management: Asset management is the business strategies that include inventory, contractual and financial functions to support management and decision making for the IIoT environment. The Assets management help to increase reliability of IIoT components assets that are including hardware and software elements. Organizations and companies have to develop effective IIoT asset management strategies to reduce the incremental risks and saving costs of industrial infrastructure.

Smart Machines: Smart machines are intelligent devices. These machines are the main element in IIoT environment that uses Machine to Machine (M2M) technology to communicate and interact with other devices in industrial.

3.3   Benefits of Industrial Internet of Things

Industrial Internet of things has many benefits. These are listed below.

1. Improve operational efficiency through predictive maintenance and remote management.
2. Performance enhancement in economy and increased visibility into marketing to produce high quality products.
3. Collaboration between people, devices and the cloud which will result in improving levels of productivity.
4. Improve performance, safety, reliability, and energy efficiency by connecting sensors to analytics systems.
5. Tracking performance for real-time marketing of products in the world.
6. Process optimization and energy consumption for control systems through analytics of big data generated from sensors.
7. Reduce maintenance time at manufacturing plants through real-time monitoring equipments.
8. Make effective market decisions based on analyzing the collected data from sensors and devices.

3.4 Security Issues in IIoT

Control systems have different types of threats and attacks associated with them so that special issues should take in consideration to secure, protect and handle these systems against malicious users. Control systems is important field in Industrial Internet of things where large number of industrial devices are connecting together through IP-supported network to serve the industry field to produce products for help people to life in better level of comfortable.

Security is very important for the safe and reliable operation of IIoT connected devices. Security aspects in IIoT will be different from classical IoT due to the difference in IIoT structure. In the recent time, devices in the control system use IP address to manage easily and handle them locally and remotely via a network. This network may connect to the Internet so that it may be attacked by malicious users and attackers. New types of malicious software and viruses are developed to breach the security of control systems like stuxnet virus. In IIoT, security consideration will differ from the IoT system where there is security triangle i.e., Confidentiality, Integrity, and Availability (CIA) which should be taken into consideration for IoT systems. But the industrial systems have another parameter which is very important and critical. This parameter is safety which is used to guarantee and protect people life.

In the past days, industrial systems [9] were not connected to the IT network so that little threats and attacks. Security is important issue for securing communication system between things in industrial system especially in the IIoT. There is still a need for developing new mechanisms and methods to protect industrial infrastructure. Many vendors made some progress in this area but still there is a serious need for more security solutions to secure the communication process in the IIoT. There are many considerations to be taken to protect the industrial Internet of Things environments as follows:

- Developing and building secure industrial control environments will help improve the growth of the industrial Internet of things.
- Incorporating industrial security into real world will give security professionals a complete view of the threat landscape, helping them better evaluatation the overall risks and threats against organizations in order to build and create better security and privacy solutions.
- Develop new policies to secure industrial environment to be more safe and reliable.
- Giving a training about security awareness for industrial system for workers and engineers who manage and operate control system to understand the risk of threats in industrial field.

3.5   *Future Research Directions in IIoT*

In the next years, organizations and companies will use Industrial Internet of Things (IIoT) to make novel growth in the industrial field. There are many open problems that are needed to be solved by providing suitable solutions and suggestions. In this section, some research points are suggested to help researchers to identify problems and gaps in IIoT technology such as follows:

1. *To develop novel big data solutions*: Big data solutions are needed to analyze data which are generated from machines and sensors. These solutions will help to predict and make decisions to increase productivity.
2. *To develop strategies and methods for energy and power management*: Power and energy management help to make industrial devices to work more time especially in critical application.
3. *To develop new security and privacy techniques*: These techniques will help to protect and secure sensitive information inside the IIoT environment.
4. *To design and develop new digital forensic procedures and techniques*: Developing digital forensic procedures and techniques will help to discover weakness in addition to trace attackers because using new techniques and tools to make crimes which are novel for industrial control system. These crimes are unique and designed for breaching the security of IIoT infrastructure.
5. *Cloud Portability*: Standards for support integration of cloud computing and IIoT to provide huge capabilities of resources will improve the productivity and performance of industrial control system.
6. *Interoperability*: Interoperability is currently the biggest challenge facing the industrial IOT so that is an important research point form researchers is to overcome this challenge.

4   Conclusion

In the recent time, advanced development in nanotechnology led to nanomachines and nano-scale devices. They communicate and interconnect via the Internet. The new networking paradigm which is called Internet of nano things have a great impact in several applications in people daily life, ranging from environmental to industrial and military. In the other side, Industrial Internet of things has drawn a significant research attention for many researchers to design and develop new techniques and methods in industrial control systems based on the use of Internet of things technology. The IIoT is considered as a part of IoT that will comprise billions of intelligent communicating industrial things together. Contents of this chapter are written to help for the understanding of concepts, communication issues, communication classifications, architectures, applications, benefits, security and future research directions which are required to achieve in depth knowledge about IoNT and IIoT technologies.

References

1. Gubbi, J., Buyya, R., Marusic, S., Palaniswami, M.: Internet of things: a vision, architectural elements, and future directions. Future Gener. Comput. Syst. **29**(7), 1645–1660 (2013)
2. Atzori, L., Iera, A., Morabito, G.: The internet of things: a survey. Comput. Netw. **54**(5), 2787–2805 (2010)
3. Li, S., Xu, L.D., Zhao, S.: The internet of things: a survey. Inf. Syst. Front. **17**(2), 243–259 (2014)
4. Akyildiz, I.F., Jornet, J.M.: The internet of nano-things. IEEE Wirel. Commun. **15**(7), 58–63 (2010)
5. Akyildiz, I.F., Pierobon, M., Balasubramanian, S., Koucheryavy, Y.: The internet of bio-nano things. IEEE Commun. Mag. **53**(3), 32–40 (2015)
6. Jornet, J.M., Akyildiz, I.F.: The internet of multimedia nano-things. Nano Commun. Netw. **3**(4), 242–251 (2012)
7. Balasubramaniam, S., Kangasharju, J.: Realizing the internet of nano things: challenges, solutions, and applications. IEEE Trans. Comput. **46**(2), 62–68 (2013)
8. Akyildiz, I.F., Jornet, J.M.: Electromagnetic wireless nanosensor networks. Nano Commun. Netw. **1**(1), 3–19 (2010)
9. Meltzer, D.: Securing the industrial internet of things. ISSA J. **1**(1), 24–30 (2015)

Part II
Distributed Systems and Security
in Internet of Things

Secure Distributed Group Rekeying Scheme for Cluster Based Wireless Sensor Networks Using Multilevel Encryption

Mohsen Yazdinejad, Faezeh Nayyeri, Omid Mahdi Ebadati E. and Nasim Afshari

Abstract Recently evolution in various devices, aid unrealistic dreams of connection between them and people on a large scale comes true and accelerates the idea of the Internet of Things (IoT). Sensors are playing an important role in this way. Conversely wireless sensor networks take the advantages to build a bridge and become an edge for IoT. By enhancing the usage of wireless sensor networks in many fields such as military, having a secure environment for communication within sensors is becoming more obligatory. Wireless sensor networks can be affected by active or passive attacks. One of the important concerns to reach to the aim of the IoT after deploying WSNs is its security. In this chapter, a model that is dealing with a variety of passive attacks, including node capturing and eavesdropping, which make data to be eliminated or changed is proposed. Due to the resource constrained of WSNs, secured schemes that use in ad hoc and wired networks are not appropriate. Additionally, to ensure environmental security and reduce usage of limited resources of sensors, a novel scheme using dynamic key management and encrypted data security is proposed. In order to prevent invader's access to encrypted data, generating a new key during re-keying model is developed. For this purpose, a random number is sent as a pattern instead of regular sending a new key to a group of sensors for encryption. Index of an array, which is pre-distributed among sensors is determined through location of 1's in the binary representation of the pattern, involve keys, which used in encryption or decryption. The results show that the complexity of time is $O(K * N)$,

M. Yazdinejad · F. Nayyeri · N. Afshari
Department of Knowledge Engineering and Decision Sciences,
Kharazmi University, Tehran, Iran
e-mail: mohsen.yazdinejad@gmail.com

F. Nayyeri
e-mail: faez.na@gmail.com

N. Afshari
e-mail: nasim.afshari@gmail.com

O.M. Ebadati E. (✉)
Department of Mathematics & Computer Science,
Kharazmi University, Tehran, Iran
e-mail: omidit@gmail.com ; ebadati@khu.ac.ir

© Springer International Publishing AG 2017
D.P. Acharjya and M. Kalaiselvi Geetha (eds.), *Internet of Things:
Novel Advances and Envisioned Applications*, Studies in Big Data 25,
DOI 10.1007/978-3-319-53472-5_6

where K is the size of the array, which its keys use in encryption or decryption, and N is the length of the text that should be encrypted or decrypted. Furthermore, by paying more time linearly the security of the network is increased exponentially for the larger scales of networks.

1 Introduction

The concept of the Internet of Things (IoT), is that everything is interconnected and accessible into a global dynamic network. Sensors and smart objects are beside classical computing devices key parties of the IoT. Since sensors are a key part of IoT, thus are wireless sensor networks (WSN) [1]. Wireless sensor networks (WSNs) consist of a larger number of small sensor nodes with limited resources such as energy, memory and computational capabilities, and some base stations (BS). BSs are much more powerful nodes that connect the sensor nodes to the rest of the networks.

Nowadays, Wireless Sensor Networks are vital for the Internet of Things (IoT). One of the fundamental features of WSNs is that the sensor nodes are expected to operate for long periods of time without attention from a higher-level presence (e.g. a human) [2]. Moreover, they are primarily designed for realtime collection and analysis of data in adversary environments. For this reasons WSNs are well suited to a large amount of monitoring and surveillance applications such as wild-life monitoring, bushfire response, military command, intelligent communications, industrial quality control, observation of critical infrastructures, smart buildings, distributed robotics, traffic monitoring, examining human heart rates, remote plant control and etc. [3, 4].

Nodes of WSNs are often placed in a hostile environment, where they are not protected. Therefore, WSNs like any wireless networks are at risk for various attacks [3]. The monitoring and spoofing of a communication channel by unauthorized attackers are known as passive attack, such as eavesdropping and capturing. Attackers to WSNs may attempt to eavesdrop on the network and capture cluster heads (CHs) or sensors to drop the messages totally or selectively, inject a fake data in aggregated results or read data from its memory. Thus, it is essential to have more security to WSNs. According to WSNs limitations, designing an effective and energy-efficient security solution in WSNs is a challenging issue [5].

In order to secure the communication between sensors and the BS, key management is required. The goal of key management in WSNs is to solve the problem of creating, distributing and maintaining those secret keys [6]. Therefore, techniques that may use in this field are very important. So, key management systems gained some interest, and various schemes have been proposed for these key management systems. However, some other researchers try to make WSNs more efficient in their energy consumption and routings. They proposed different models to reach to an optimize energy model of a sensor [7], segmentation of the sensors area [8] and managing the routing protocols of the sensors [9].

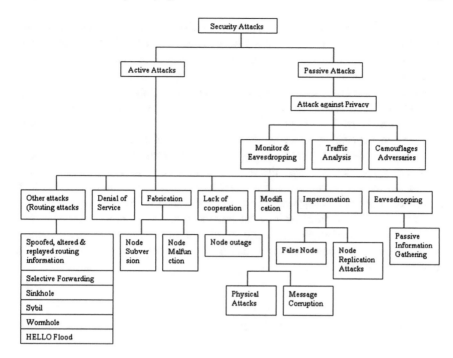

Fig. 1 General classification of security attacks

Most of the prior works on securing sensor data use traditional security solutions that are based on cryptographic algorithms such as SPINS, or LEAP [10]. In general, to stand against invaders, hash function and re-keying are used. Figure 1 shows the classification of various attacks categories called as active and passive. On the other hand, the key management for WSNs can generally categorize into static keying and dynamic keying based on whether the keys after the initial usage of sensor networks are updated or not, which is often termed as re-keying. Re-keying occurred when a node is captured, which would compromise the keys stored in that particular node. Thus, one advantage of a dynamic key management scheme is to ensure that the network is not captured by an attacker, and it is the replacement of old keys that is discovered by invaders.

Typically, grouping is a technique to do localize computation and reduce communication overhead in WSNs. Furthermore, clustering is the most common technique in grouping. However, the other advantage of clustering is that if a node is compromised in a cluster, the re-keying process is implemented only for that group (group re-keying). Many researchers have been interested in group re-keyings due to its low computational and decrease of complexity [11–15].

In order to implement access control, and prevention of capturing the network, [16], evaluate the problem of adding security to cluster based communication protocols for similar wireless sensor networks, consisting of sensor nodes with severely limited resources and proposed a security solution for LEACH. Despite several solutions that provided in recent work for security improvement of WSNs, attackers can reach to new keys, which is sent during the re-keying time, and they could change the information of keys.

The aim of this chapter is to introduce a method to improve the Internet of Things WSNs security and protect it from these kinds of attacks by creating a new secret key for sensors and CHs. For this purpose, a novel efficient re-keying mechanism is used to implement a special patterned key. In the proposed model, on the stage of re-keying instead of sending a new key to a group of captured sensors, a numerical pattern has been sent to the sensors. Then, the sensors encrypt data according to the received pattern and key arrays, which pre-distributed to them. The encryption method, which sensors used, is affine encryption. The rest of the chapter is arranged as follows; next section is a review literature on the related work, after that the proposed method and the security model for wireless sensor networks are presented, next is the experiment and analysis of the proposed method and finally the last part is the conclusion and future of this work.

2 Literature Review and Objectives

Presently, researchers are working on different innovations and techniques to integrate WSN better into the IoT environment [1]. Network security and its related technologies become one of the important key challenges in recent years. Consequently, various research and experimental schemes have been proposed. A secure confidence routing mechanism by using network-based intrusion detection system is proposed [17], and to continue of this research, they have proposed chasing intruders by the aim of mobile agents [18]. A dynamic rule-based approach based firewall policy [19] to implement distributed firewall or hybrid method with the aid of honeypots showed a significant defense method to the known and unknown attacks.

New strategies and technical reliability of current securities need best practice and professional investigation. Optimization of micro-firewalls as a distributed method to lay in critical infrastructure became a priority to regain the control and latest assessment for large enterprises [20, 21]. However, WSNs and its security provided other broad range issues, including different vulnerabilities, intrusion detections, secure routing, encryption and cryptography, and aims by various applications. Turkanovic et al. [22] proposed a highly efficient and novel user authentication and key agreement scheme (UAKAS) for heterogeneous WSN (HWSN), which was adapted to the IoT notion. Moreover, their scheme is considerably efficient since it is built on a simple symmetric cryptosystem. Sabzinejad et al. [1] found that the former method of Turkanovic has some security deficiencies and is prone to some cryptographic assaults. So, they focus on overcoming the security weaknesses of that scheme, by proposing a new and improved UAKAS. The proposed method enables the same functionality, but increases the security level and enables the HWSN to dynamically raise without influencing any party involved in the UAKAS.

One of the advantages of WSNs is dynamic in the sense that they allow the addition and deletion of sensor nodes after deployment, to grow the network or replace failing of unreliable nodes. As mentioned before, WSNs are at risk to various security assaults such as eavesdropping (overhear message transmissions), jamming (interfere with or block message reception) and node capturing attacks (steal all the information stored within the captured node) [4].

The necessary cryptographic primitives for WSNs are block ciphers, message authentication codes (MACs) and hash function [23]. The already available hash functions (such as SHA1 or MD5) are relatively cheap and can be used in sensor nodes without a significant overhead. In addition, message authentication codes can be constructed from block ciphers [24]. Singaravelu and Verma, [25] investigated the scheme used multivariate cryptosystem. In that scheme, the private key includes the two invertible affine transformations, which makes encryption much faster than some schemes. In addition, Komninos, Soroush, and Salajegheh [23] introduced the design and implementation of light-weight energy-efficient encryption algorithm (LEE) that can be used in a tiny constrained device, especially sensor nodes to provide security services for the communications.

Another technique to save both computation and communication resources of WSNs, is data aggregation. One of the threats that injected a fake data in aggregated results is the sensors to be captured. Therefore, data have to be secured during aggregation using cryptographic techniques. While existing solutions cause a significant overhead, Boudia et al. [26] present a novel secure data aggregation scheme for WSNs by using Stateful Public Key Cryptography (SASPKC) to provide an efficient security. This scheme can achieve a high security level with a low overhead in computation and communication in large-scale scenario. Besides, according to [27], there are three types of cryptographic techniques for the WSNs has been developed: symmetric, asymmetric, and hybrid. Since WSNs has limited constrains, the more lightweight techniques like the symmetric ones are more suitable, because symmetric based cryptographic schemes required less computational power.

Turkanovic et al. [28] proposed a novel user mutual authentication key agreement scheme for heterogeneous WSNs. Then, they have developed a scheme based on symmetric cryptography. It has been claimed that their scheme in comparison to the more similar ones, is less computationally cost and has more security features. But this scheme requires more storage space than others-related ones. Nevertheless, [29] demonstrated several security weaknesses of the Turkanovic et al. protocol. In order to fix the security pitfalls, they have designed a novel architecture, which has been presented for user authentication and key agreement scheme for the WSNs environment. Their protocol, which not only resists the security weaknesses, but also achieves complete security requirements.

In eavesdropping scope, most of the current work has only focused on mitigating its activities [30–32] or protecting the communications by using encryption algorithms [33]. However, a novel analytical framework to eavesdropping attacks and model is in wireless networks, has been proposed by Li et al. [34]. This proposed framework enables to evaluate the impact of various factors on the eavesdropping attacks theoretically. Furthermore, it can calculate the probability of eavesdropping attacks, depending on the environments. Then they pro-posed how to reduce the probability of eavesdropping. Also, Bonaci et al. [4] investigated eavesdropping attacks in WSNs, that can be prevented by making use of symmetric key based cryptographic techniques to control access to communication among sensor nodes. Conversely, the use of cryptography requires a mechanism to generate, distribute and when needed, revoke and refreshes cryptographic keys used for secure communication.

Capturing a node is another hazard that threats the sensors. It enables attackers not only to hold of cryptographic keys and protocol states, but also to clone and redeploy malicious nodes in the network [4]. This threat allows the attackers to extract and spoof data from their memories. As pointed out in [35], the existing adversarial models such as Byzantine failure or Dolev–Yao threat models are inadequate to describe node capture attacks. In [36], Eschenauer and Gligor present a key pre-distribution method that allows secure deployment of WSNs. They also presented node revocation, and keys update techniques to ensure network connectivity and services. Tague and Poovendran, [2] investigated the problem of modeling node capture attacks in WSNs in terms of the impact of the attacks on the network protocols, and security by decomposing it into a primitive set of events. In this work, they presented a model for node capture attacks in WSNs, incorporating attack evaluation metrics into a framework for optimal attacks. Bonaci et al. [4] present such a comprehensive framework that is simultaneously able to integrate secure deployment techniques, corrupted nodes identification algorithms and node revocation methods in order to study WSNs resilience to node capture attacks. They derive a linear dynamical model for a WSN under node capture attacks and different controllers in order to control the network response to node capture attacks. Figure 2 presents the WSN attack classification.

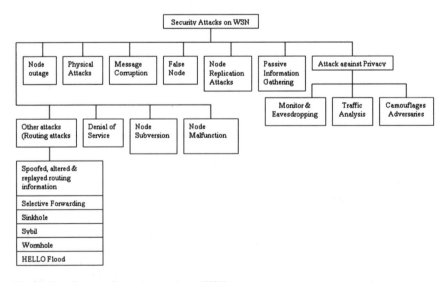

Fig. 2 Classification of security attacks on WSN

Key management is a core mechanism to ensure security in network services, applications of WSNs and communication between sensors and the BS. As it has been already mentioned before, according to wide application of WSNs, one of the fundamental security issues that involved many researchers, industrial engineers and schemes interests are working on dynamic key management. He et al. [6] presented an overview of dynamic key management schemes in WSNs. This chapter investigated security, advantages and disadvantages of each presented schema.

It has to be mentioned that it is not possible to find one single perfect scheme that can perform well in all evaluation metrics, as each of them has some definite strengths, weaknesses and suitability for specific situations. In any dynamic key management scheme, the re-keying mechanism can be done either periodically or on demand in response to a node capture. Thus, one of the advantages of a dynamic key management scheme is, the replacement of old keys that are revealed to an attacker in order to ensure that the network is not captured by them [14].

Grouping is a technique to do localize computation and to reduce communication overhead in WSNs. Dini and Savino [12] have proposed a group key revocation protocol for WSNs. This protocol also provides a lightweight key authentication using one-way hash chains. The authors have shown that using this structure, the numbers of messages are reduced, but some of the messages should be sent to more than one member. But the drawback of this scheme is that, it does not provide backward secrecy, and that is due to the use of hash chains of keys. Jiang et al. [13], offered a scheme that improves the Dini and Savino works by using a dual hash chain, in which nodes can authenticate the new group key, but the revocation could not be completed on-demand, and it should be planned in advanced. Also, there is an implicit assumption that the attacker is not able to compromise group nodes, hence, could not read the keying materials in their memory.

Another protocol for updating group key has been proposed in [14] by Khan et al.. They adapt the secret-sharing revocation scheme by reducing the computational overhead. Zhang in [11], has proposed a group re-keying protocol. In this protocol each member distributes encrypted shares of its re-keying materials to other nodes, which will be returned to the node in a re-keying event. To determine the efficiency of any re-keying scheme for WSNs depends on several parameters, Khan et al. [14] in their paper presented an efficient re-keying mechanism for cluster-based WSN, which requires broadcast of only one message from the base station for re-keying.

Seyed et al. [15] proposed a new group re-keying scheme, which is based on local collaboration of group members. It can be used with any key size. The re-keying process is handled locally in the group itself. Wang et al. [37] investigated an efficient re-keying mechanism, which employs a Lagrange polynomial interpolation to calculate the new group key, that also scalable and is based on secure protocol for key revocation in WSNs. The protocol guarantees an authenticated distribution of new keys that is proficient in terms of storage, communication and computing overhead. The proposed protocol reduces the number and the size of re-keying messages. However, according to Eschenauer and Gligor, traditional key pre-distribution offers inadequate solutions that each being pair-wise privately shared with another node, and must be installed on every sensor node. Eschenauer and Gligor [36] proposed a simple key pre-distribution scheme that requires memory storage, and yet has similar security and superior operational properties in comparison to those of the pairwise private keys-sharing scheme. Their scheme relies on key sharing among the nodes, uses a simple shared-key discovery protocol for key distribution, revocation and node re-keying.

3 Proposed Model

Wireless sensor networks are subjected to many attacks due to their wide range of applications. Therefore, security is an important issue to this subject. It has been mentioned before that commonly, attacks are classified as the active and passive category, however, monitoring and spoofing by unauthorized is coming under passive's class. On the other hand, the attacks against privacy are passive in nature. Therefore, the proposed model in this study focuses on variety of passive attacks, including, eavesdropping and sensor capturing. Management of secured key for all available nodes in a network during attacking would be having a high overhead of computational, communication and energy consumption. Therefore, grouping technique is deployed for the sensors within the networks. Clustering is the most common technique in grouping. One of the advantages of clustering is, if a node is captured in a cluster, the re-keying procedure is applied only for that cluster, which is under attacks. Figure 3 presents the same activity on a particular node.

In this chapter, a new model of re-keying that immune the WSNs from passive attacks is proposed. This model is suitable in various industrial environments including, actuators, smart meters, pole-top devices, wireless body sensor networks and new smart home appliances, along with that various other usages are included different military, aerospace and transportation purposes. For assuring that compromised

Fig. 3 Re-keying procedure is applied only for the cluster, which are under attacks

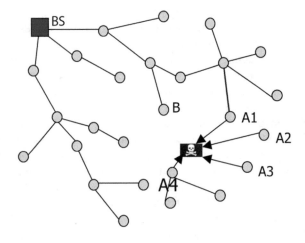

nodes do not take part in the group communication. It is assumed that group members share a secret pattern that uses to authenticate messages, which is already broadcast within the group. Upon detecting a compromised sensor node, a new group pattern key must be distributed to all that group members have a secured network. On the other hand, due to a limited lifetime, cryptographic keys in WSNs can expire. Therefore, patterned keys used for secure communication in WSNs and should be periodically updated. For applying security in a network to deal with passive attacks, which is captured sensors, a string that is included numbers, will be sent to the sensors before, instead of the key. Then, each CHs send a random key to all their member groups to determine that which numbers in a pre-distributed array must be used in cryptography. Moreover, in order to avoiding of capturing of the whole network, when the attackers capture a sensor, CH generates a random number again and then, informs all other sensors in the group under attack except compromised sensor (the re-keying process that already has been described will initiate).

In order to improve security of the network, multi-level cryptography is deployed. The sensor encrypts data, according to location of 1's in the binary representation of sending random number in the pre-distributed array. Further, affine cipher would be used for each selected key in the pre-distributed array. This means that, at first the first key encodes the sent message, and then the second key encodes the result of the first step, etc. It has to be mentioned that to provide a complete security for the network, the re-keying would be performed for all clusters, periodically. Details of the proposed model is explained in the following sections.

3.1 Affine Cryptography

Affine encryption is one of the well-known cryptosystem in the substitution of ciphers. Actually, in case of the substitution cipher encryption/decryption are arithmetical operations, but it is more convenient to think of encryption as permutations of characters and so the decryption function will be the inverse permutation. The special aim of the substitution of the cipher with the affine cipher is restricting the encryption functions to functions in the form of:

$$Encrypt(x) \equiv (Ax + B) \ mod \ M \tag{1}$$

where A and B are the secret keys to cipher, so that number will be chosen from integers $\{0, 1, 2, 3, \cdots, (M-1)\}$, and x is the symbol to be encrypted, and modulus M is the endpoint size to choose.

In affine cipher, each plaintext is first mapped to the integers in the range [0, M−1] for a fixed integer M. Then it uses modular arithmetic to transform the integer into another integer named cipher-text. Note that, A and M are co-prime, so it can be decrypted, where the decryption function is defined below.

$$Decrypt(x) \equiv A^{-1}(x - B)(mod \ M) \tag{2}$$

where A^{-1} is the modular multiplicative inverse of A modulo M, i.e., it satisfies the equation:

$$AA^{-1} \equiv 1(mod\ M) \tag{3}$$

It can be shown that, decryption function is the inverse of the encryption function as defined in Eq. 4. Hence, forward each key means a pair of A and B.

$$
\begin{aligned}
Decrypt(Encrypt(x)) &\equiv A^{-1}(E(x) - B)(mod\ M) \\
&\equiv A^{-1}(Ax + B(mod\ M) - B)(mod\ M) \\
&\equiv A^{-1}(Ax + B - B)(mod\ M) \\
&\equiv x\ (mod\ M)
\end{aligned}
\tag{4}
$$

3.2 Mechanism of Selection of Keys from an Array

As aforementioned, in our re-keying phase, CHs send a random key to sensors to determine that which numbers in the pre-distributed array must be selected to use in encryption. In this section, its functionality is described in details.

1. At first, in the re-keying process, a number between 1 and 2^{K-1} will be generated by CH and send to sensors.
2. Positions of 1's in the binary representation of the generated number by CH, determine indexes of the array, which their keys involve in encryption. Therefore, the sensors detect the positions of 1's of this number and then extract the keys.

In fact, in this model re-keying means generating another random number in the range of $[1, 2^{K-1}]$. Each number in this range, represents a unique combination of the keys. Each sensor receives a number, which is sent by CH as a pattern, then selects the keys from the pre-distributed array (keys []) and saves them in a new array (selectedKeys []) by using the following Algorithm 1. After that the keys in the selectedKeys[] will be used in the encryption.

In this algorithm, index i is for keys[] and index j is for SelectedKeys[]. In fact, i^{th} index moves from the rightmost bit toward the leftmost bit of pattern. The pattern which is sent by CH will be shown by P. If the rightmost bit of P equals to 1 that means it's odd. So the first number in the keys' array should be selected and if the P is even, then the first number in this array will not be selected. The rightmost bit of P is being dismissed by dividing it by 2.

Algorithm 1 *(Selection of Keys)*

1. $i = 0$
2. $j = 0$
3. *while (pattern $!= 0$)*
4.   *if (pattern % 2 == 1)*
5.   *selectedKeys[j] = keys[i]*
6. $j = j + 1$
7. $i = i + 1$
8.   *pattern = pattern/2*

The mentioned process will do for new numbers till pattern (P) will be 0. In every executing of "while loop", i increases by one and a digit is being dismissed from the right side of the pattern. At the end of this algorithm, j keys have selected from *keys*[ ] and have stored in *selectKeys*[ ]. In other words, *keys[i]* is selected when *pattern* & $2^i \neq 0$. Otherwise, it is not selected. The index i varies from 0 to K, i.e., $0 \leq i < K$ and & is the bitwise AND operator. K is the size of the array that its keys use in encryption or decryption.

Example 1 Suppose that $K = 10$ and CH sends the number 621 as a pattern P to the sensors. The following steps will be done to detect the keys in the pre-distributed array. These selected keys are ready for using in the encryption process.

$621\%2 = 1$ (621 is odd), so *Keys*[0] is selected.
pattern = pattern/2, and the new value of the pattern is $621/2 = 310$.
$310\%2 \neq 1$, and so *Keys*[1] is not selected.
pattern $= 310/2 = 155$
$155\%2 = 1$, and so *Keys*[2] is selected.
pattern $= 155/2 = 77$
$77\%2 = 1$, so *Keys*[3] is selected.
pattern $= 77/2 = 38$
$38\%2 \neq 1$, and so *Keys*[4] is not selected.
pattern $= 38/2 = 19$
$19\%2 = 1$, so *Keys*[5] is selected.
pattern $= 19/2 = 9$
$9\%2 = 1$, so *Keys*[6] is selected.
pattern $= 9/2 = 4$
$4\%2 \neq 1$, so *Keys*[7] is not selected.
pattern $= 4/2 = 2$
$2\%2 \neq 1$, so *Keys*[8] is not selected.
pattern $= 2/2 = 1$
$1\%2 = 1$, so *Keys*[9] is selected.
pattern $= 1/2 = 0$

Fig. 4 Positions of 1's in
binary representation of 621

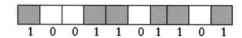

1	0	0	1	1	0	1	1	0	1

In Fig. 4, the binary representation of the number 621 has been shown. Green cells represent the indexes of the pre-distributed array that their keys involve in encryption. According to the above illustration, each message which is sent from CH to the sensors, at first encodes by *Keys*[0] then its result encoded by *Keys*[2] and so on.

The re-keying process in CH can be done with any of random number generator algorithms or built-in functions in programming languages. However, it should mention that the consumption time is over ten times running of the algorithm and for each time the average consumed time of encryption/decryption is presented.

3.2.1 Generating Controlled Random Numbers

It is obvious that the complexity of encryption is depending on selectedKeys[] (number of 1's in the binary representation of random number). Therefore, the fewer number of 1's in random number has the lower complexity. Generation of random numbers can be controlled in a way that number of 1's are not scanty (because of security) and are not numerous because of complexity of time. This section, introduces an algorithm for generating random number with control on the number of 1's. Controlling the number of 1's in the binary representation of the random number can be done by setting upper and lower bounds for the number of 1's.

Algorithm 2 *(Controlled Random Number Generation)*

1. *$A[K] = \{0, 1, 2, \ldots, (K - 1)\}$*
2. *oneCount = random $[(K/2) - 2, (K/2) + 2)]$*
3. *$A = random - shuffle(A)$*
4. *mask = 0*
5. *for $i = 0$ to oneCount − 1*
6.   *mask = mask + $(2^{A[i]})$*
7. *return mask*

Random number must be selected neither great to increase time complexity, nor very small for endangerment security. So, the number of 1's should be determined randomly from a range of $[(K/2) - 2, (K/2) + 2]$ (oneCount). In fact, lower bound has been considered as $(K/2) - 2$ and upper bound as $(K/2) + 2$. Then, a random permutation from 0 to $(K - 1)$ is generated and selected the first oneCount's numbers from the permutation. These numbers will be the positions of 1's in the binary representation of the random number. The following pseudo code expresses the process of generating random numbers.

Example 2 Consider $K = 10$ and after executing line 3, *oneCount* $= 6$ and $A = \{6, 2, 0, 9, 3, 5, 8, 1, 7, 4\}$ and by running the loop, the result of masking will be as below:

mask $= 0 + 2^6 = 0 + 64 = 64$
mask $= 64 + 2^2 = 64 + 4 = 68$
mask $= 68 + 2^0 = 68 + 1 = 69$
mask $= 69 + 2^9 = 69 + 512 = 581$
mask $= 581 + 2^3 = 581 + 8 = 589$
mask $= 589 + 2^5 = 589 + 32 = 621$

Therefore, the output of function in this case will be 621. This function should be located in each sensor, which is potentially capable of being CH. So, if it detects sensor capturing in the group, it generates another random number and sends it to the other compromised sensors. As it is presented, the re-keying process by CH, includes generation another random number in the range of $[1, 2^{K-1}]$ (of course while it controls the number of 1's). The Simulation of random number generation has been done by using C++.

4 Experimental Work

To deal with attackers that are trying to capture sensors over time, re-keying process is being done. If encryption and decryption with the same previous key can be done, the attacker who captured a sensor could easily decrypt the data by receiving new messages, and after encryption can send any message to the CH. Besides, to deal with the attackers who eavesdrop on the network, a number is being sent to the sensors instead of sending a new key in the re-keying process. Therefore, the sensors can perform encryption by using the numbers and the material that they already have. These kinds of attackers, can eavesdrop on new keys in re-keying in addition to eavesdropping the encrypted text (cipher text). Hence, they can use the key and encrypt the texts, nevertheless, by implantation of the proposed method; the attacker will not be able to understand the method of encryption/decryption even with the eavesdropping the numbers that have been sent to the sensors in the re-keying process.

4.1 Analysis

However, using proposed method assured the security against two types of attackers, but the complexity will increase in terms of time and memory. As the K gets larger, the number of different combinations that can be made with keys, increases. Therefore, there is a tradeoff between security and complexity. As it is already presented by

Table 1 Memory consumption (kilobyte), running time (ms) and number of instructions (million) of the proposed mechanism for a text with the length: 50000

K	Pattern	Number of 1's	Memory (KB)	Time (ms)	Instruction (MI)
10	37	3	288.7	136	60.14
	610	4	288.7	175	76.52
	398	5	288.7	218	91.96
	821	6	288.7	257	108.6
	727	7	288.7	297	125.0
	7846	8	288.7	336	141.7
	44211	9	288.7	379	158.6
20	828589	10	288.7	414	174.6
	402173	11	288.7	463	190.6
	841275	12	288.7	504	205.9
		Simple Affine Encryption	273.4	46	27.41

rising of K, the number of various combinations that can be made with keys, also grows exponentially that aids to increase security exponentially too, while the growth of complexity of time is linear and memory is constant. So, it seems that the K can be increased as much as the memory, energy of sensors and respond time endorsed in which, sensors are allowed to respond to the CH.

Presented multi-level encryption has been simulated with the C++ language program that was implemented on Intel core i5 with 6 G.B of RAM, and running at a clock frequency of 1.7 MHz running Windows and Fedora operating system for various analyzers and simulators, included, Massif-visualizer, Valgrind and Matlab. There are three inputs defined in this simulation: a Boolean, an integer, and a string. The Boolean is for determination of encryption or decryption, the integer is the pattern that is shown by P, and the string is the text that should be encrypted or decrypted depended on the flag. If the flag is true, the text will be encrypted by P, otherwise, the text will be decrypted by P and K. Then, by using the predefined keys in the program encryption will execute according to the algorithm that already explained. In Table 1, the running time of the program is shown in a text with 50000 characters, where $K = 10$ and $K = 20$ with consideration of the number of 1's in pattern.

4.2 Memory Consumption

Depend upon the sensor type, memory requirement is decided for implementing the proposed method. There might need a little quantity of extra memory required as it is used to store the programs for decoding the pattern and for random number generation. As it has been mentioned before, each sensor which is capable of being

Fig. 5 Memory
consumption based on the
proposed model

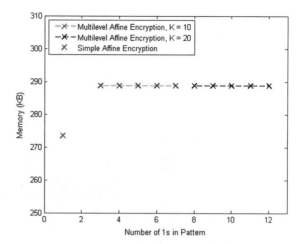

CH should have a random number generation program. Meanwhile, providing keys
to the sensors requires memory, which depends on the K and type of keys.

It's worth noting that the selected Keys array is written for a better understanding
of the concept of pseudocode and it is not used in the implementation process. In
fact, encryption will execute with corresponding key, while each position of 1's in
the pattern determines the level of encryption. Therefore, the number of 1's in the
pattern has no effect on the quantity of storage memory (Fig. 5). Table 1, presents the
memory consumption, running time and number of instructions of the program in
a text with 50000 characters, where $K = 10$ and $K = 20$ with consideration of the
number of 1's in pattern.

These measurements have been done by Valgrind that is an instrumentation framework for building dynamic analysis tools. It gives a few snapshots and information
about memory usage during the run time of program and stores them. In order to
compare a pattern with variety of 1, Figs. 6, 7, and 8, illustrate the memory consumption, by implementing massif-visualizer. In these figures memory consumption
in encryption with a minimum number of 1's and $K = 10$ (Fig. 6), maximum number
of 1 s and $K = 20$ (Fig. 7), and simple affine encryption (Fig. 8) is depicted. It has
to be mentioned that in the present scheme, the number of 1 s in pattern represents
the number of encryption that must be done. In the following figures, the X-axis
represents time, and the Y-axis represents how much memory is consumed.

4.3 Time Consumption

In re-keying phase, each sensor has to extract keys from pre-distributed array and
encrypt data with them. So in fact the total complexity of each sensor is calculated
by using Eq. 5.

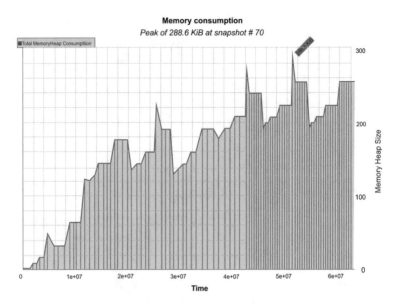

Fig. 6 Visualization of the proposed model on (pattern = 37) by massif-visualizer tool

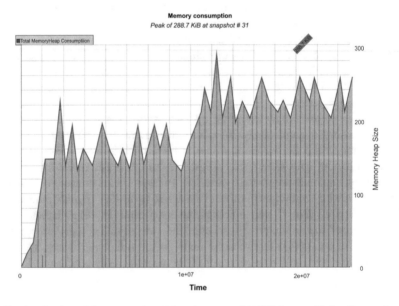

Fig. 7 Visualization of the proposed model on (pattern = 841275) by massif-visualizer tool

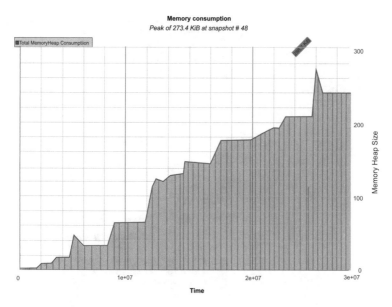

Fig. 8 Visualization of simple affine encryption by the proposed method (considering Table 1) by massif-visualizer tool

$$\text{Total Complexity} = \text{Complexity of Encryption} + \text{Complexity of Extracting Keys} \tag{5}$$

Complexity of encryption in this phase will be presented with $O(N)$, such that N is the length of the text that should be encrypted or decrypted. Assuming M is the number of 1's in pattern, then the time complexity of encryption is $O(M * N)$. Complexity of extracting keys is $O(lgP)$ or $O(K)$ where P is the sent number toward sensors by CH and K is the size of the array that its keys use in encryption. It is obvious that the complexity of encryption is depending on the size of *selectedKeys*[ ] (number of 1's in the binary representation of random number). However, the fewer number of 1's in random number has the lower complexity.

On the other side, as it has mentioned before, generation of random numbers can be controlled in a way that number of 1's are not scanty and numerous. This function should be located in each sensor, which is potentially capable of being CH. CH will generate random numbers in the re-keying process. The time complexity of this function depends on two parts: first part is related to random shuffle, and the second is related to the loop that the complexity of both of them is O(K). So, the total complexity for generating random numbers is $O(K)$. Therefore, overhead time in this phase is related to extracting keys, multi-level encryption and generating random numbers (for CHs). So, the total time complexity of re-keying phase, in the worst case, is $O(K + M * N)$.

Since in the proposed model, the number of 1's is nearly K/2, $O(M) = O(K)$. Thus, $O(K + M * N) = O(K + K * N)$. Consequently,

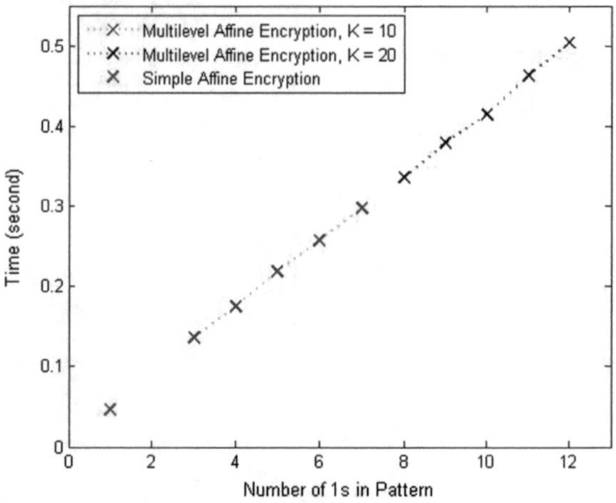

Fig. 9 Comparison of the
running time of the proposed
mechanisms for $K = 10$ and
$K = 20$ with consideration
of the number of 1's in
pattern, and simple affine
encryption

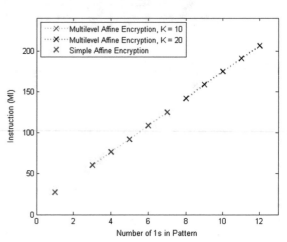

Fig. 10 Comparison of the
number of instructions of the
proposed mechanisms for
$K = 10$ and $K = 20$ with
consideration of the number
of 1's in pattern, and simple
affine encryption

$$Complexity\ of\ time = O(K * N) \tag{6}$$

Figure 9 illustrates the comparison of running time of the proposed mechanism for $K = 10$ and $K = 20$ with consideration of the number of 1 s in pattern, and running a simple affine encryption. Of course, it is clear that running time is depending on the number of instructions, which are executed. Since, the number of 1 s in pattern represents the number of encryption, which must be done, as shown in Fig. 10, increasing the number of 1 s in pattern will cause increase instruction that reason increases the time.

Finally, in Fig. 11 all factors of memory, running time, number of instruction and security of the proposed model are shown based on K. As it is clear, with increasing of K, the security of our model increases exponentially (2K) while the number of instructions and time also is increasing linearly, and memory is nearly constant.

Fig. 11 Comparison of all factors (memory consumption, running time, number of instruction and security of the proposed model. (PCK: possible combination keys)

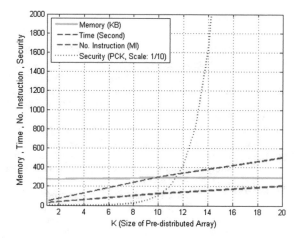

5   Conclusion and Future Work

Adapting wireless sensor networks in the next generation of products is one of the aims of the IoT to connect all four angles of this square, which are processes, people, data and things. Based on this idea security of the sensors is becoming one of the major issues. This chapter proposed a new re-keying mechanism for WSN, where there might face a passive attack such as eavesdropping and capturing of a sensor which, make data to be eliminated or changed in wireless sensor networks. The model is applicable to use for various IoT sensors, wireless body sensor networks, smart home appliances, military, aerospace and transportation. Due to the resource-constrained of WSNs, secured schemes that use in this network is critical. To ensure the security, in presenting scheme, CHs send "one random number" to the nodes instead of sending a new key, then multi-level encryption is done with this number for sending secured and authenticated messages between sensors and CH in a group. The proposed model is implemented with C++ in the visual studio ver. 2013 and then compared the outputs based on levels of encryption, which is dependent upon the number of 1's in random number sent by CH. The results show that by paying more time linearly and without sensitive change in memory consumption, the security of the network is increased exponentially for the larger scales of networks, however, increasing the time is negligible as for the number of 841275, it will be only 504ms and much better than the current available methods. The future work includes optimization of this mechanism for a less time-consuming for larger scale of networks.

References

1. Sabzinejad Farash, M., Turkanovic, M., Kumari, S., Holbl, M.: An efficient user authentication and key agreement scheme for heterogeneous wireless sensor network tailored for the internet of things environment. Ad Hoc Netw. **36**(1), 152–176 (2016)
2. Tague, P., Poovendran, R.: Modeling node capture attacks in wireless sensor networks. In: Proceedings of 46th Annual Allerton Conference on Communication, Control, and Computing, vol. 3, pp. 205–209 (2008)
3. Padmavathi, D.G., Shanmugapriya, M.: A survey of attacks, security mechanisms and challenges in wireless sensor networks. Technical Report on arXiv, preprint arXiv:0909.0576 (2009)
4. Bonaci, T., Bushnell, L., Poovendran, R.: Node capture attacks in wireless sensor networks: a system theoretic approach. In: Proceedings of 49th IEEE Conference on Decision and Control, vol. 4, pp. 67–72 (2010)
5. Masdari, M.S., Bidaki, M.: Analysis of secure LEACH-based clustering protocols in wireless sensor networks. Wirel. Netw. **9**(3), 345–358 (2013)
6. He, X., Niedermeier, M., Meer, D.H.: Dynamic key management in wireless sensor networks: a survey. J. Netw. Comput. Appl. **36**(2), 611–622 (2013)
7. Yaeghoobi, S.B.K., Ebadati, O.M.E.: Performance of triangle segments tree model to optimize energy efficiency in wireless sensor networks. In: Proceedings of 3rd International Conference on Machine Learning and Computing, vol. 3, pp. 314–320 (2011)
8. Yaeghoobi, S.B.K., Soni, M.K., Tyagi, S.S., Ebadati, E.O.M.: Impact of NP-complete in triangle segments tree energy efficiency model in wireless sensor networks. J. Basic Appl. Sci. Result **3**(9), 808–817 (2013)
9. Yaeghoobi, S.B.K., Tyagi, S.S., Soni, M.K., Ebadati, E.O.M.: SAERP: an energy efficiency real-time routing protocol in WSNs. Proc. Int. Conf. Optim. Reliab. Inf. Technol. **5**, 617–622 (2014)
10. Kamel, I.H.J.: A lightweight data integrity scheme for sensor networks. Wirel. Sens. Netw. **4**(2), 213–221 (2011)
11. Zhang, W.G.C.: Group rekeying for filtering false data in sensor networks: a predistribution and local collaborationbased approach. J. Sens. **3**(5), 213–229 (2005)
12. Dini, G., Savino, I.M.: An efficient key revocation protocol for wireless sensor networks. Proc. Int. Symp. IEEE Comput. Soc. World Wirel Mob. Multimed. Netw. **4**, 213–219 (2010)
13. Jiang, Y., Chuang, L., Minghui, S., Xuemin, S.: Self-healing group key distribution with time-limited node revocation for wireless sensor networks. Ad Hoc Netw. **5**(1), 14–23 (2007)
14. Khan, F.I., Jameel, H., Raazi, S.M.K., Khan, A.M., Huh, E.M.: An efficient re-keying scheme for cluster based wireless sensor networks. In: Proceedings of International Conference on Computational Science and Its Applications, pp. 1028–1037 (2007)
15. Seyed, H., Amir, H., Vanesa, D.: A distributed group rekeying scheme for wireless sensor networks. In: Proceedings of the 6th International Conference on Systems and Networks Communications, vol. 4, pp. 345–349 (2011)
16. Ferreira, A.C., Vilaa, M.A., Oliveira, B.L., Habib, E., Wong, H.C., Loureiro, A.A.: On the security of cluster-based communication protocols for wireless sensor networks. In: Proceedings of International Conference on Networking, pp. 449–458 (2005)
17. Ebadati, E.O.M., Kaur, H., Alam, M.A.: A secure confidence routing mechanism by using network-based intrusion detection systems over router monitoring controls. OLS J. Wirel. Inf. Netw. Bus. Inf. Syst. **6**(1), 1–11 (2010)
18. Ebadati, E.O.M., Kaur, H., Alam, M.A.: A performance analysis of chasing intruders by implementing mobile agents. Int. J. Secur. **4**(4), 80–88 (2010)
19. Kaur, H., Ebadati, E.O.M., Alam, M.A.: Implementation of portion approach in distributed firewall application for network security framework. Int. J. Comput. Sci. Issues **8**(6), 211–221 (2011)
20. Kaur, H., Ebadati, E.O.M., Alam, M.A.: Optimization of micro-firewalls security layers in distributed network. In: Proceedings of International Conference on Future Information Technology, pp. 312–320. IACSIT Press (2011)

21. Ebadati, E.O.M., Kaur, H., Alam, M.A., Yaeghoobi, K.: Micro-firewalls for dynamic network security framework: a secure hybrid architecture. J. Netw. Secur. **4**(2), 315–324 (2012)
22. Turkanovic, M., Brumen, B., Holbl, M.: A novel user authentication and key agreementscheme for heterogeneous ad hoc wireless sensor networks based on the internet of things notion. Ad Hoc Netw. **20**(2), 96–112 (2014)
23. Komninos, N., Soroush, H., Salajegheh, M.: Light weight energy efficient encryption algorithm for sensor networks. In: Proceedings of IEEE 9th International Symposium on Communication Theory and Applications, pp. 312–317 (2007)
24. Preneel, B.V.R.: Cryptographic primitives for information authentication state of the art. J. Cryptogr. **2**(3), 45–54 (1998)
25. Singaravelu, P., Verma, S.: Performance analysis of multivariate cryptosystem schemes for wireless sensor network. Comput. Electr. Eng. **39**(6), 1880–1893 (2013)
26. Boudia, O.R.M., Senouci, S.M., Feham, M.: A novel secure aggregation scheme for wireless sensor networks using stateful public key cryptography. Ad Hoc Netw. **4**(2), 34–45 (2015)
27. Sharma, G., Bala, S., Verma, A.K.: Security frameworks for wireless sensor networks-review. Procedia Technol. **6**(1), 978–987 (2012)
28. Turkanovi, M., Brumen, B., Hlbl, M.: A novel user authentication and key agreement scheme for heterogeneous ad hoc wireless sensor networks based on the internet of things notion. Ad Hoc Netw. **20**(3), 96–112 (2014)
29. Amin, R., Biswas, G.: A secure light weight scheme for user authentication and key agreement in multi-gateway based wireless sensor networks. Ad Hoc Netw. **23**(2), 67–81 (2015)
30. Lu, X., Lio, P., Wicker, F.: Security estimation model with directional antennas. In: Proceedings of IEEE International Conference on Military Communications, pp. 234–239 (2008)
31. Wang, Q., Dai, H.N., Zhao, Q.: Eavesdropping security in wireless ad hoc networks with directional antennas. In: Proceedings of 22nd IEEE Iinternational Conference on Wireless and Optical Communication Conference, vol. 4, pp. 221–229 (2013)
32. Dai, H.-N., Wang, Q., Li, D., Wong, C.-W.R.: On eavesdropping attacks in wireless sensor networks with directional antennas. Int. J. Distrib. Sens. Netw. **5**(2), 67–78 (2013)
33. Zafer, M., Agrawal, D., Srivatsa, M.: Limitations of generating a secret key using wireless fading under active adversary. IEEE/ACM Trans. Netw. **20**(5), 1440–1451 (2012)
34. Li, X., Xu, J., Hong-Ning, D., Qinglin, Z., Chak Fong, C., Qiu, W.: On modeling eavesdropping attacks in wireless networks. J. Comput. Sci. **15**(2), 67–81 (2014)
35. Parno, A.P.B., Gligor, V.D.: Distributed detection of node replication attacks in sensor networks. J. Secur. Sens. Netw. **8**(4), 56–67 (2005)
36. Eschenauer, L., Gligor, V.D.: A key-management scheme for distributed sensor networks. In: Proceedings of the 9th ACM conference on Computer and communications security, pp. 313–318 (2002)
37. Wang, Y., Ramamurthy, B., Zou, X.: An efficient key revocation scheme for wireless sensor networks. In: Proceedings of IEEE International Conference on Communications, pp. 78–83 (2007)

Recognizing Attacks in Wireless Sensor Network in View of Internet of Things

D.P. Acharjya and N. Syed Siraj Ahmed

Abstract Wireless sensor networks are wide spreading, dominating, and has a far-reaching range of utilizations such as battlefield scrutiny, traffic scrutiny, forest fire detection, flood detection etc. Both industry and academia are targeting their research works for the sake of advancing their functions. The safety of a wireless sensor network is negotiated due to the random distribution of sensor nodes in exposed environment, memory restraints, power restraints and unattended nature. Furthermore, providing confidence between every couple of communicating nodes is a demanding issue and is of prime importance in internet of things. Under these conditions this chapter spotlights variety of attacks and their symptoms thoroughly. The chapter analyzes them based on trusting based protection techniques including classical techniques such as fuzzy, Bayesian, game theory etc. Modern techniques such as clustering, bio-inspired computing, key establishment based techniques are stressed upon to provide maximum protection for each node in the wireless sensor network.

1 Introduction

Currently sensors are accessible everywhere. Wireless sensor networks have become one of the most attractive and promising field over the past few years. We take it for granted, but there are sensors in our smart phones, vehicles, ground monitoring soil conditions in vineyards and even in the factories controlling CO_2 emissions. Although it seems that sensors have been around for a while, the study on wireless sensor networks initiated back in the 1980s. But, recently wireless sensor networks produced an increased interest because of industrial perspectives. It is because of the availability of cheaper, tiny in size low powered miniature components such as

D.P. Acharjya (✉) · N.S.S. Ahmed
VIT University, Vellore, Tamilnadu, India
e-mail: dpacharjya@gmail.com

N.S.S. Ahmed
e-mail: nssa.26@gmail.com

© Springer International Publishing AG 2017
D.P. Acharjya and M. Kalaiselvi Geetha (eds.), *Internet of Things: Novel Advances and Envisioned Applications*, Studies in Big Data 25,
DOI 10.1007/978-3-319-53472-5_7

149

radios, sensor, and processors that were often unified on a single chip to monitor some particular phenomenon collectively. Akyildiz et al., proposes the traits such as flexibility, self-organization, high sensing fidelity, fault tolerance, cheaper and rapid distribution. It created many new and current applications in real life such as wildlife monitoring, disaster response, military scrutiny, smart building, industrial quality control, battlefield scrutiny, traffic scrutiny, forest fire detection, flood detection, humidity recording, temperature recording, pressure monitoring and light monitoring inside the area of distribution [1]. The recorded measurements are sent back to base station. The transmission range of sensor nodes are restricted to tens of meters and hence not all of them can directly transmit with the base station. Therefore, information is sent hop-by-hop from one sensor node to another node until they reach the base station.

Protection becomes an important issue in wireless sensor networks and brings new threats for protection engineers. Karlof et al., discusses various kinds of attacks against wireless sensor networks have been detected so far such as sinkhole attack, sensed data attack, black hole attack, bogus routing, selective forwarding attack, wormhole attack, and hello flood attack, etc. [2]. All the protection solutions introduced so far can be partitioned into two models: prevention based models and detection based models. Prevention based models, such as encryption and certification, are often regarded as the first line of defense against attacks. Further more cryptographic models can also be used to prevent an external attacker from altering the ongoing transmission. The field of distribution is not usually physically protected and an attacker can easily enter into the field, and imprisonment some nodes. Since they are not tamper-resistant, the attacker can modify the software running on the nodes to launch an array of internal attacks such as selective forwarding, jamming, and hello flood attacks. Detection based models are designed to identify and detach internal attackers after failing of prevention based model. However, the harm caused by internal attacks might be serious.

Intrusion detection systems are proper tools to defend against internal attacks and they are generally used in common networks. Furthermore, there are two kinds of intrusion detection systems: signature based detection and anomaly based detection. Signature based detection systems compare the known attack profiles with suspicious behaviors whereas anomaly based detection systems recognize unusual deviations from pre-established common profiles to identify the abnormal behaviors. However, these intrusion detection systems cannot be directly applied to wireless sensor networks mainly due to severe restraints of sensor nodes on energy, memory and computing power. It is enough to have several probes in common networks if they are fixed in traffic assembly points. In wireless sensor networks, some attacks can be identified only by the neighbors of a malicious node. Hence, we assume that each sensor node runs an intrusion detection system agent and supervises its neighbors in indiscriminate mode. It is believed that the collected information is analyzed locally by a sensor node itself or in collectively with only nodes from the close surroundings. This is because of high energy consuming transmission, where one transmitted bit consumes the power of executing 800 instructions [3].

This chapter analyzes various attacks, protection requirements, trust analysis and challenges pertaining to neighborhood based anomaly detection model [4, 5]. Following the introduction in Sect. 1, symptoms of various attacks in wireless sensor network is presented in Sect. 2. Protection requirements in wireless sensor networks is discussed in Sect. 3 followed by analysis of trust in Sect. 4. Challenges to trust techniques in wireless sensor networks is presented in Sect. 5. A brief about modern trust techniques for wireless sensor network is offered in Sect. 6 followed by conclusion in Sect. 7.

2   Symptoms of Various Attacks in Wireless Sensor Network

This section describe selective forwarding, hello flood, jamming, routing anomalies and identity duplication including packet alteration attack together with their symptoms which can be applied in the neighborhood based model to identify these attacks. These symptoms help us in identifying the concerned attack.

2.1   Jamming

In jamming attack a malicious node intently tries to intervene with physical transmission and reception of wireless connection as shown in Fig. 1. This chapter, discuss various jamming attacks such as deceptive, constant, random, and reactive. A deceptive jammer uniformly injects regular packets without any gap between subsequent

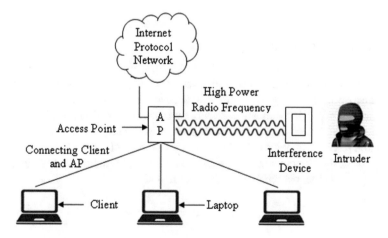

Fig. 1 Diagrammatic representation of jamming attack

packet transmissions. A random jammer substitutes between sleeping and jamming randomly. A reactive jammer stays muted when the channel is idle but starts transmitting as soon as it senses action on the channel [5]. Generally, jamming attack deals with signal strength, packet sending and receiving rate, packet delivery ratio, and certify rate.

Signal strength (SS) is an analysis of the power present in a received wireless signal. The signal strength of the node n_{ij} received at a node is denoted as $SS(n_{ij})$. The circulation of $SS(n_{ij})$ is affected by a deceptive jammer. A node in the network considers the node n_{ij} malicious if the circulation of $SS(n_{ij})$ significantly differs from circulations of $\{SS(n_{i1}), \cdots , SS(n_{im_i})\}$. This approach of jamming cannot be applied to reactive and random jammers [5].

Packet sending rate (PS) is the amount of packets sent over a fixed period of time. The packet sending rate of the node n_{ij} noticed by a node is denoted as $PS(n_{ij})$. This help in identifying deceptive jammer. Packet sending rate of a deceptive jammer significantly distinct from the neighboring nodes since it sends an abnormal rate of packets in comparison to legitimate nodes. On the other hand, packet receiving rate is the total amount of packets received over a fixed period of time. It is of two types such as packet receiving rate with and without retransmissions. The combination of packet receiving rate with the packet sending rate can be used to identify a random jammer.

Packet delivery ratio is the ratio of packets that are favorably delivered to a sink compared to the number of packets that were sent by a sender. It can be computed as the ratio of the number of received certified packets with respect to the number of sent packets. This approach is oftenly applied in reactive jammer. Similarly, the packet certify rate is the amount of certifies (acknowledgements) sent to the node that are noticed by a node. Existence of any type of jammer in a nodes neighborhood can be noticed based on packet delivery ratio drops down [5]. But, identification of attacker itself is not disclosed.

2.2 Hello Flood Attack

In hello flood attack, a malicious node circulates hello packets using a more powerful transceiver than a general sensor node. Nodes receiving such hello packets may falsely assume that they are within the wireless range of the sender and try to advancing their packets through this malicious node. These packets will not reach its destination [2]. This attack can be detected with the help of received signal strength. In such cases, the signal strength of nearest neighbor of a malicious node is significantly higher than signal received from other neighbors.

2.3 Selective Forwarding

Packets are oftenly transmitted from one node to other in a network. Packet forwarding rate of a specific node is the number of packets that the node received from its neighbors and appropriately advanced to its destination node during a fixed period of time. By default, in many cases packets are dropped during transmission and based on this packet dropping rate is computed. Packet dropping rate is the number of packets that were sent to a particular node but are not advanced by that node. A node having considerably higher packet dropping rate than other nodes in the neighborhood is treated as malicious. A malicious node frequently refuses to advance certain packets and simply drops them. It leads to an attack called as selective forwarding. Such type of attacks can be detected by observing the packet dropping rate of neighbouring nodes. The sensor nodes that are neighbor to each other should have very much identical packet dropping rates. If any part of a network is congested or affected by environmental changes then majority of nodes in that part will have the packet dropping rates similar.

2.4 Routing Anomalies

Routing anomalies are more delicate forms of faults. In such cases, a faulty node cooperate in the information gathering process, but it does not do so according to the protocol requirement by design. The impact of which is generally local and restraints to individual nodes. A routing anomaly often heads to coordinate faulty behaviors among multiple nodes. A node may only advance a subset of traffic and mistreat some. Routing anomaly is produced due to a number of reasons such as buggy routing module, injection of false routing message, and malicious routing attack from foreign entities.

The prime complexity is how to accomplish software faults methodologically. One solution is to utilize fault reports from current sensor network projects whereas the other is to correct programs at random places. Unfortunately, the first solution is not useful as these faults are not well recorded. But, faults created this way may not be very significant. In contrast, considerable work has been performed on routing attacks and their defenses in reference [7]. Various well recorded routing attacks to produce significant routing faults are black hole, sink hole, flooding, warm hole, grey hole etc. This section talk about such attacks briefly along with its symptoms.

2.4.1 Black Hole

In wireless communication, black holes refer to areas in the network where arriving or advancing traffic is wordlessly discarded without conveying the source that the information has not reached its intended recipient. When analyzing the topology

Fig. 2 Diagrammatic
representation of *black hole*
attack

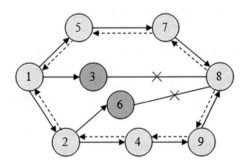

of the network, the black holes themselves are hidden, and can only be identified by noticing the lost traffic. In multi-hop wireless sensor networks, nodes act as routers to circulate messages from their children to their corresponding parents and eventually to the sink. A black hole node drops the sending packets from its child nodes. Compared to fail-stop breakdowns, the malicious node may appear normal to its parent as it can still transmit its local information. A graphical presentation is presented in Fig. 2. In Fig. 2, it is clearly seen that node 3 and node 6 are malicious.

Additionally in black hole attack, the sender node receive reply message from fault node and make shortest way to receiver node. Malicious node sends reply message after authorised node to sender node and then sender become confuse in two replies. In this, the data packets fully dropped by sender node. The sender node sends large amount of route request message to every nearby nodes. When route request message is received by malicious node, then it sends route reply message to sender node which is un-real and also shows the shortest path to reach to receiver node. Sender node accepts the reply message from un-real node which is termed malicious node and transfers the packets.

Black hole attack are of two types such as internal black hole attack and external black hole attack. In internal black hole attack, an internal node become the fault node and makes route from sender node to receiver node. In external black hole attack, attack occurs from outside of the network. It is called as denial of service (DoS) attack. In this attack, network take advantage from network traffic and collapse the whole network. It is done by an external fault node and then working as same as internal node.

2.4.2 Sink Hole

The sinkhole attack is specifically a serious attack that blocks the base station from acquiring complete and accurate sensing information. Consequently, it produces serious threats to higher-layer applications. In this attack, a malicious node tries to draw complete or as much traffic as possible from a particular place, by making itself looking attractive to the neighborhood nodes with respect to the routing measured. As a result, the attacker succeeds to attract all traffic that is directed to the base station. By taking part in the routing process and generating more serious attacks like selective forward, modify or even packet drop [2].

A black hole restraints by the number of child nodes along with the sink tree, which use the malicious node as relay. In contrast, a sink hole is a more serious form of black hole. A malicious node communicates fault routing information often by advertising a router of shorter hops to the sink. As a result more neighborhood nodes attract to use this node as relay. Simultaneously, more nodes are disconnected from the sink when the malicious node drops all the transit traffic.

2.4.3 Flooding

Flooding is an attempt to block the victim from being able to use complete or part of wireless connection. Flooding attack allows an attacker to pervert, subvert, disrupt, or destroy a network, and also to minimize a networks capacity to provide a service [8, 9]. Additionally, such attack widen to all the layers of the protocols of stack. They are generally very complicated to block because they exist in various forms inside the network. For instance, a malicious node can send large number of requests to a server which has to test the legality of the nodes. Due to the large amount of requests, the server will be busy in testing illegal requests and so it will not be available for the legal ones. This motivates declining the performance of the entire network as the network gets blocked due to illegal requests.

2.4.4 Worm Hole

This type of attack occurs on the routing protocol in which the packets or single bits of the packets are acquired at one location, channeled to another location and then replayed at some other location [10, 11]. In this attack two or more attackers are connected by high speed off channel link called wormhole link. Attackers forms

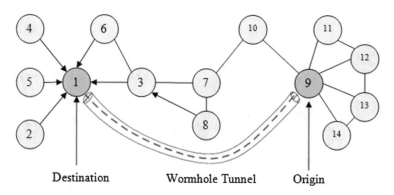

Fig. 3 Representation of wormhole attack

tunnels to transfer the data packets and replays them into the network. This attack has a tremendous effect specially against routing protocols. Routing mechanisms are generally confused and disrupted because control messages are tunneled to wrong direction. The tunnel created between the two attackers is termed as wormhole link. Figure 3 depicts the working of wormhole attack. Packets received by node 1 is replayed through node 9 and vice versa. In general, it take several hops for a packet to traverse from a location near node 1 to a location near node 9. Packets transmitted near node 1 travelling through the wormhole will arrive at node 9 before packets travelling through multiple hops in the network. The attacker can make other nodes believe that they are neighbors by forwarding routing messages. Further, the attacker selectively drop data messages to disrupt communication between other nodes in the network.

2.4.5 Gray Hole

Gray hole is an alternation of black hole attack in which the nodes drops packet selectively. In this attack a legitimate node works as nasty nodes and shows itself normal. It takes place in the reception of the packets transmitted from the source node, on reception of packets these nasty nodes drops the selected packets and only transmits the left packets to the neighbor node. These nodes behave as usual and therefore this attack is complicated to detect. Simultaneously, a gray hole attack influence one or two nodes in the network. The various ways in which a nesty node can drop packets are given below.

1. It can drop complete UDP packets while transmitting complete TCP packets.
2. It can drop 50–60% of the packets or can drop them by some probability distribution.

Figure 4 depicts gray hole attack. In the figure, node 7 act as gray hole. Gray hole attack can be classified according to node count in wireless sensor network such as single nasty node, two consecutive nasty node, non consecutive nasty node, and surrounding nasty node.

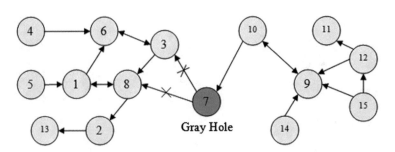

Fig. 4 Representation of *gray hole* attack

3   Protection Requirements in Wireless Sensor Networks

It is believed that, the diversity and the difficulty of attacks have extended to many extents. Therefore, it is essential to think of protection to put a stop to the attacks. This will help to decrease the harmful causes of similar warnings. Protection systems are expressed to a set of protection necessities. These necessities depend on few fundamental protection characteristics and few definite protection designs in a built-in source restraint of wireless sensor networks. Brief discussion on these concepts are presented in the subsequent sections.

3.1   Protection Characteristics

Lopez et al. has fully outlined protection characteristics and current protection systems in wireless sensor networks. Authors also consider the following points would usually be piece of the protection necessities in wireless sensor networks [13]. These are listed below:

1. Authentication: Authentication of messages permits a recipient to cross validate the message which is actually send through the stated correspondent.
2. Confidentiality: Any specified information should not be disclosed by any other than the specific receivers.
3. Integrity: The message created and inspired through the sensor network should not be falsely modified.
4. Availability: The nodes of a wireless sensor network should be able to receive its services each and every time they require them.
5. Authorization: Restricted to authorized individuals to perform selective operations in the network.
6. Freshness: The message created by the wireless sensor network should be fresh.
7. Self organization: Nodes of sensor should be self determining and supple sufficient to separately react next to challenging circumstances, and be capable to arrange and repair themselves.
8. Forward and backward confidentiality: Forward confidentiality is when a sensor must not be able to study any prospect information following it leaves from the network. Backward confidentiality is when a unification of sensor must not be capable to study any formerly communicated information.
9. Non-repudiation: A node can not be refused to send information that it have.
10. Auditing: The fundamental component of a sensor network should be capable to back up any important actions that happen within the network.
11. Privacy and anonymity: The place and characteristics of the base class, and the nodes that produce message must be kept secret or protected.

3.2 Protection Systems

Protection system of wireless sensor networks has been suggested by various researchers [13–16]. The overview of current protection systems include the followings:

1. Protection primitives: The protection primitives for wireless sensor networks offer the privacy, reliability, validation, and non-repudiation characteristics. It contains keys of asymmetric, symmetric cryptography, functions of hash, and information validation code.
2. Key management and protected channels: The objective of key managing techniques is to resolve the issue of circulating, generating, and keeping of confidential keys. Alternatively it leads to form confidential paths. The key managing techniques are divided into four chief systems such as negotiation system, key-bunch system, mathematical system, and asymmetric key system. This partition is done to cover several diverse categories of protocols.
3. Network core protocols: It deals with routing, message collection and moment coordination. The properties and behavior of these protocols are extremely vulnerable on the trait of wireless sensor network function wherever they are executing. It should be suitable to the demands of the defacto system.
4. Self-restorative and self-managing protocols: Self-alertness mechanism provides data of a specific node which leave from a region to the sensor node through protocols. Self-repairing techniques are used to assist the formation of protection functions like intrusion detection systems and trust managing systems.
5. Privacy and anonymity: Because of a specific situation, the secrecy of the network components is to be made in explanation. The menaces faced by secrecy come mostly as of content, place and uniqueness.
6. Software based safeguard and testing: Perverted or duplicated nodes can be detected as of legal nodes using on specific software support models like isolated verification and wireless fingerprinting.
7. Supplementary protocols and techniques for a particular wireless sensor networks: Besides the core protocols, there are supplementary protocols and services that can be helpful to recover the performance of the network like cluster management. In addition, remaining aspects of protection in wireless sensor networks that require additional investigation are: the association among the protection and quality of service necessities of the system, techniques like to attack tree functional to wireless sensor network to measure their menaces, evolution of safe place designs, formulation of secure architectures and middle-wares that use cross-layer optimizations, investigation of name and addressing vulnerabilities, information redundancy and survivability, distributed computing, tools for the defense of the medium access control layer etc.

As well Hoffman et al. analyzed the protection tools engaged using current standing systems and proposed five protection designs next to attackers for benign methods in general networks [17]. These are briefly explained below:

1. Blocking multiform identification (Sybil attacks): Clarifications to contract with attacks is projected into streamline and circulate ways. In streamline way, a vital authority delivers and cross checks recommendation of each individual uniquely. Conversely, circulate way do not depend on a vital individual. Several answers are projected including binding a unique identification or with network correlates to detect nodes with various identifications.
2. Mollifying the production of fake rumors: Several accesses prevent the production of fake rumors using digital signatures and certain evidences to incorporate responsibility.
3. Mollifying the diffusion of fake rumors: Two systems has been introduced to decrease the diffusion and collection of fake influences. The former relying on trusting of pre-defined identifications. But it is extra dangerous if it is malicious. The later is used arithmetical methods like Bayesian principles to form an actual comment system, defining a certain threshold for abnormal behaviors.
4. Blocking the misuse of short range model: To restrict attackers misusing the model using hiring their influence decrease rapidly and again introduce into the model with a novel identification. One way to make certain that fresh comers should start with a little influence for a fixed period of time.
5. Mollifying the denial of service attacks: To block a denial of service attack next to extending, one way is used to incidental models to assemble members for computation and widening of influence principles. This may partition accountability for either whole identifications or decrease the possibility of cooperativeness as of compromising nodes.

4 Analysis of Trust in Wireless Sensor Networks

Trust method is one of the protection mechanism in which insider attacks are prevented from varied and constitute a self-restorative wireless sensor network. The advancement trend of trust is branched into two directions: authorization or hard trust and evaluation or soft trust. Blaze et al. first introduced a trust management system, called method maker. It is introduced to define and enforce protection method, certificate and connection that permit direct authorization of critical protection actions [18]. Inherent ideas of trust are managed by applications which are established on cryptographic models. A trusted middle man signs a message certificate to authorize the identity associated with a public key. An important problem in this scenario is how the authorized identity is behaved upon. The reaction to this is just a authorization policy of trust. In particular, the trust management system certainly implements the process of finding whether access should be permitted according to the defined policy, authorization semantics and access rights.

Chapin et al. have examined the foundations and characteristics of authorization in trust management [19]. But, the prime requirement is how to compute the range of trust. This issue is consequently discussed in many literature, providing distinct trust partition, computation models in various domains such as e-commerce, peer-to-peer, adhoc networks etc. [20]. Various fundamental properties of trust in wireless sensor networks, including definitions of trust, characteristics of trust and trust values are stressed upon in this section. Several methodologies for trust systems in common networks, which could be applicable to wireless sensor networks after directed modification are also discussed. Furthermore, due to the implicit defined characteristics of wireless sensor networks, the individuality of trust mechanisms matched with other networks are specified.

4.1 Characteristics of Trust

Issues of trust definition in various schemes of wireless sensor networks is discussed by much researchers. Momani et al. introduced information trust and communication trust [21], while Lin and Varadharajan suggest hybrid trust based on soft trust and hard trust [22]. As well other researchers targets on node trust, path trust, and service trust by listening the behaviors of nodes, the connectedness of paths, and the connections of service respectively. Even though there is no clear solution on the trust definitions in wireless sensor networks, much of them are concluded that, trust is a biased opinion in the loyalty of other entities or functions, including fairness of information, connectedness of path, handling capability of node, connections of service etc.

Based up on the statements of trust properties in the literature, trust has common characteristics in different networks. It is planned below.

1. Subjective: This is supported by some observers or recommenders, relying upon some records of preceding behaviors.
2. Reflexive: Undoubtedly, it trusts itself.
3. Asymmetric: It is commonly independent between two sides. It means to say, node P trusts node Q while node Q may distrust node P.
4. Incomplete transitive: The trust link builds, but relying upon the structure or range of the trust relationships among neighborhoods. It means that, node P trusts node Q, and node Q trusts node R, while node P may trust or distrust node R.
5. Context sensitive: It is effective within a specified pre-requisite.
6. Dynamic: It may vary by period and space.

In addition to above characteristics, there are three more current special attributes of trust in wireless sensor networks upon the above-listed information about network environments is introduced.

1. Low evidence: The trust evidence including self behaviors or approved from others, is rarely poised due to the intrinsic characters of wireless sensor networks like restraints energy power or uncovered network environment.
2. Ambiguous instability: The variation of dependability status may be large, which is damaged easily by the wireless environment itself, however, it can not be blamed on the attacker.

3. Complicated trust system structure: Similar to the diverse structures of wireless sensor networks, there are different architectures of trust systems: centralized trust systems, decentralized or distributed trust systems and hybrid trust systems. These rely upon the scores of trust calculations.

4.2 Trust Computations

In the previous subsection, characteristics of trust are discussed. Based on the characteristics, trust computations or relationships can be partitioned into three types such as direct trust, recommended trust, and indirect trust.

1. Direct trust: Direct trust is evaluated by noticing the one hop neighbor directly and making sufficient investigation, their trust relationships are established without depending on intermediaries.
2. Recommended trust: Recommended trust is evaluated based on recommendation given by a third party about a particular node. It is designed based on pull cumulative and push cumulative. In pull cumulative based design a trust demanding node will need to find the trust information providing on the sink node and then pull the trust information and various cumulative trust values from various providers. In the push cumulative based design each member node in the network who has evaluated the trust of the sink node and also ready to share the evaluated trust information, push its evaluated trust value on sink node to other members in the network. Any member can collect its interested trust information, and cumulate them on various propagation path.
3. Indirect trust: Indirect trust is acquired when communicating entities verify the validity of each other members trust based on the authorization given by the centralized server. It also engages in communications like direct trust methods however here the communications requires authentication and verifications from the authentication server. Moreover, it is an integration of both direct trust and recommended trust.

4.3 Trust Values

The mode of trust value may not be crucial for general networks. However for the wireless sensor networks, the implication is completely different. In common, documenting and evaluating real quantities will need much more time and storage complexity. In other words it takes more energy level and high storage space. But, it is to be avoided in wireless sensor networks. Thus, this problem is essential due to the form of trust value. But, where to collect those trust information, and how to progress and calculate trust information in logical and realistic terms is of prime concern.

Trust values provide different methods of calculation of trust. It is proposed two types of definitions for computation of trust values such as continuous and discrete. Earlier, there are variable ranges of trust according to the concrete calculation method. For instance, the common range could be $[-1, 1]$. Later it got changed to an integer quantity or discrete value with tags rather than sizes [23]. Furthermore, the preset trust values are divided into three types as discussed below.

1. High starting score: When the network is safe at the commencement of constructions, it may be adopted high starting score in an encouraged manner.
2. Low starting score: When the network is not safe at the commencement, it would be better to be doubtful and have low starting score, holding a discouraged manner.
3. Middle starting score: Here the ranking of trust for all members are identical at the commencement, with expression of indifferent manner.

4.4 Classical Trusting Techniques in Wireless Sensor Networks

Over the past several decades, much researchers have suggested trusting or influence systems in different fields, which spread from electronic commerce to information technology [24–26]. Few consequent trusting techniques for the wireless sensor networks are introduced in favor of gaining a wide understanding. For this, it is essential to explain some of the trusting techniques with several classical mathematical measures. The following subsection provides information briefly about these classical measures.

4.4.1 Fuzzy Trusting Technique

Self trusting is a vague relation in most cases. So, it cannot be rigidly considered with the likeliness of probability, although the value of probability varies from 0 to 1. However, fuzzy inference is a design of multi-valued inference derived from fuzzy set theory to accord with reasoning, which is almost accurate rather than accurate. Furthermore, vagueness is one of the most implicit characters in trusting networks due to the evidence backed may be fuzzy, or the methods to be required may be fuzzy. The subject inference trusting technique consists of vagueness calculation. However not all vagueness can be considered as a probability and thus cannot be expressed by a probability model. Various fuzzy trusting techniques have been suggested to support a series of fuzzy rules to manage such vagueness in trusting management. This is reputable to be applied in systems of control for decision building and pattern matching etc.

Fuzzy inference includes a series of If-Then rules. The important steps followed in fuzzy rule-based inference are listed below [27].

1. Initially formalize the fuzzy sets and criteria.
2. Assign the fuzzy engine with variable input values, on due evaluating the grade to which the fundamental input steps, and clauses of the fuzzy rules.
3. Implement the fuzzy rules to decide the output data, by evaluating the rules decision upon its matching grade.
4. Compute the outcomes and supply some comments to balance the criteria or rules.

4.4.2 Bayesian Trusting Technique

Bayesian theory is often used for trust management. It is further advanced and widely used for decades [21, 28–30]. It is categorized into two parts: objective and subjective. In objective, the statistical study on trust is carried out completely on the information study, refusing any subjective conclusion. In subjective, the confidence level is used as an argument to participate in the conclusion. In common, the trusting system schemes rely upon the past behavior records of an objects or the suitable behavior records of alternative participants. Moreover, the value of trusting can be taken as guessing of vagueness in behaviors. Luckily, the Bayesian theory is fully cooperating with the method of trusting calculation. It uses the event of prior probability, which is later revised in the light of revised suitable proofs, to create an event of posterior inference. Therefore, Bayesian theory is most relevant to classify trust value. It has been further advanced to Dirichlet distribution or β distribution [21] and united with other areas like maximum entropy [28] and Dampster-Shafer theory of evidences [30].

β **Distribution System**: Josang introduced lists of trust and influence techniques for e-commerce using beta distribution by designing a posterior influence value evaluated by binary classification inputs such as positive and negative [30]. To exhibit and obtain the influence rate, the β probability density function is applied, which is specified as the prediction value of the same. The β probability density function designated by $\beta(q|\mu, \gamma)$ and can be denoted using the gamma function τ as

$$\beta(q|\mu, \gamma) = \frac{\tau(\mu + \gamma)}{\tau(\mu) \cdot \tau(\gamma)} q^{\mu-1}(1 - q)^{\gamma-1} \tag{1}$$

where $0 \leq q \leq 1$, $\mu, \gamma > 0$ with the limitation that the probability variable $q \neq 0$ if $\mu < 1$ and $q \neq 1$ if $\gamma < 1$. The prediction value of the beta probability density function is acquired by

$$E(q) = \frac{\mu}{\mu + \gamma} \tag{2}$$

The parameters μ and γ produce the grades of p positive and m negative results with $\mu = p + 1$ and $\beta = m + 1$ respectively. After analyzing the conclusions of events, the influence grade (i.e., the prediction value of the beta probability distribution function) can be delivered.

4.4.3 Game Theory Trusting Technique

This theory aims to capture mathematical behavior in crucial situations in which an individuals progress to create a conclusion rely upon the behaviors of neighbors. To some scope, it is also known as trust game for two players [31]. In the networks of self organizing like adhoc networks, selfish behavior is a specific issue due to absence of collaboration among wide ranging nodes. Many researchers [32–34] suggest trust schemes with game theory to remove the non collaborative nodes. According to Prisoners dilemma, the inter communication operations like forwarding of packets between nodes, distribution of certain resources can be properly designed as games. However, this theory can not be taken as a tool for prediction of nodes behavior. This theory can propose how to behave members while inter communication operations. Furthermore, the necessity to implement this theory is to utilize the behavior of two-way inter communication, but it is usually one-way communication in wireless sensor networks. Hence, it is concluded that, this theory is not a suitable one to resolve trusting problems in wireless sensor networks.

4.4.4 Subjective Inference Trusting Technique

Subjective inference is expended from the Dampster-Shafer theory of evidence. It is applied for designing corrupted network and to study a Bayesian network. Subjective belief directs subjective confidence about the validity of premises with amount of vagueness [30].

A subjective inference belief is commonly represented as φ_χ^B, where B is the confidence owner also known as subject, and χ is the premises to which the belief applies. φ_χ^B is specified as a four tuple (a, b, c, d) where a represents confidence that the defined premises is true (belief), b represents confidence that the defined premises is false (disbelief), c represents degree of non-committed belief (vagueness), and d represents priori probability in lake of proof (base rate).

These elements clarify $a, b, c \in [0, 1]$ and $(a + b + c) = 1$. The prediction probability value of belief is specified as $E = (a + cd)$. The parameter d decides how vagueness is involved in the prediction value E. It is also known as relative atomicity. The subjective inference based on β distribution is defined as below.

$$E(q) = \frac{\mu}{\mu + \gamma} \tag{3}$$

where $\mu = \frac{2a}{c} + 2d$, and $\gamma = \frac{2b}{c} + 2(1 - d)$.

4.4.5 Entropy Trusting Technique

It is a theory of numerical mechanics developed from thermodynamics, and theory of information as a unit of vagueness. Caticha and Giffin suggests that Bayesian theory and largest entropy are fully suitable and can be considered as distinct way of computing largest entropy [35]. Sun et al. suggested a trusting calculation system in adhoc networks with identical method based on the Bayesian based trusting proposal design and entropy based trusting value [28]. The entropy based trusting value is specified as below.

$$E = F(q) - 1; \quad 0 \leq q < 0.5 \tag{4}$$

$$E = 1 - F(q); \quad 0.5 \leq q \leq 1 \tag{5}$$

where $E = E\{subject : agent; action\}, q = q\{subject : agent; action\}, F(q) = -q log_2^q - (1 - q) log_2^{(1-q)}$ and F is the function of entropy. The best aspect of this specification is that the trusting value is the probability of a non linear function. When the probability rises from 0.50 to 0.5095, in the former case the trusting value is incremented by 0.00024. When the probability rises from 0.990 to 0.9999 in the later case, the trusting value is incremented by 0.072. In other words whenever the vagueness is larger, i.e., the probability value is closer to 0.5., the fluctuation of trusting value is smaller.

5 Trusting Technique Challenges in Wireless Sensor Networks

The different challenges of wireless sensor networks in accordance with trusting techniques are listed below. These challenges are due to low complicatedness of algorithm, balanced period of trusting validity, diversification in the roles, and criteria estimation.

1. Low complicatedness of algorithm: Because of restraints energy level and storage level, the capacities of computation of wireless sensor nodes are low. Therefore, the design of trusting scheme for wireless sensor networks should be simple to calculate.
2. Balanced period of trusting validity: In wireless sensor networks trusting evidence is collected rarely, the trusting validity status of a wireless sensor node should be balanced. It should not be too long or too short. The long period may increases the exposure of attacks from malicious nodes, whereas short period may misuse resources from wireless sensor nodes which are at present in short supply. Hence it is necessary to provide a suitable time period in wireless sensor networks.
3. Diversification in the roles: The complicatedness structure of a trusting system needs diversification roles. It helps to meet the consistency of trusting calculation schemes. For instance, the chief node of a cluster is usually having with maximum trusting levels in the lengthy term as the base. At the same time the remaining nodes of a cluster are having minimum trusting levels and can be convicted from neighborhood information.

4. Criteria estimation for trusting system: Many trusting techniques in various direction are suggested with different analytical procedures. However, currently there is a lack of criteria consistency to evaluate the ability of those trusting techniques. Mostly simulation comparison outcomes are limited to the various parameters of the trusting technique itself. Sometimes it capture into two directions such as with trusting technique or without trusting technique. Hence, it is difficult to access a sequence of definite estimation techniques for the trusting techniques [32, 36].

6   Modern Trusting Techniques for Wireless Sensor Network

Research in computational intelligence and its applications to real life problems has been carried out in many directions. This section briefly discuss some of the modern trusting based techniques for wireless sensor networks.

6.1   Ant Colony Trusting Influence Technique

Different kinds of wireless sensor networks can be created based on type of nodes assured. One can accommodate from a static wireless sensor network where nodes having a specific location, to a fully mobile wireless sensor network where nodes can move to all places. Additionally, it can also accommodate from a very limited wireless sensor network where each and every node spends maximum time in an idle state, to other where every node supports high functional features and able to process various services per second. For simplicity it is assumed a design where a wireless sensor network is a collection of nodes with almost high sensor energy. In addition to that certain nodes demand generic operations and certain nodes support them. Also each node will communicate only to its neighboring nodes, and the nodes do not have the knowledge of complete topology of the network. Further this topology is treated as the almost highly changing. Design is made to support a node demanding some service from the network. A node is treated unreliable if it purposely supplies a fake service or supplies a wrong information because of hardware failures or performance decline. One technique that satisfies quite well for this situation is an ant colony system.

   In the literature, bio-inspired trust and influenced model for wireless sensor networks to obtain the most reliable path which leads to the most influenced node in a wireless sensor network contributing certain service is discussed. It is established fact that ant colony system cannot be straight away used effectively because of the certain constraints. Therefore, certain transformations have to be applied. Let us assume each node keeps aroma trace of its neighbors. This aroma traces $\tau \in [0, 1]$ to evaluate the probability of ants selecting some route or other. Moreover, the searching values $\eta \in [0, 1]$ are specified as the converse of the suspension communication period between any two nodes. The reality is that each node supervises its own aroma traces and searching values, but it can adjust them and becomes a great protection threat. Alternative problems that prevents the straight forward function of the ant colony system is the fact of the environment in which an ant is searching for most influenced server supplying a requested service. This could be due to inaccessible to certain nodes of ant from which a path is formed either due to switch off or due to move out of the scope of their preceding sensor in the path. The ant would not able to return back to the end user in that direction and it would get vanished. Therefore, it has no assurance that entire of them are going to return back and also it cannot pause until all the activated ants came back in one cycle of the algorithm.

The initial variation is that the main loop is specified by a general action, which may be some amount of cycles or it can even be some timeout. This definition will rely upon the particular wireless sensor network. The following algorithm provides above discussion and contains four steps.

Algorithm 3 (*Ant Colony System*)

1. Each ant appends the first sensor to its result, which is generally an end user sending from it. Then every ant elects next sensor to move according to the transition rule and it is delivered there.
2. Once each ant moved from the end user, it pauses until they return back. For each returned ant, the end user matches its result and holds the best one. As discussed before, in a wireless sensor network the end user has no assurance that all ants that were sent are going to return back, so it just pauses until a timeout is over or some percentage of ants has returned back.
3. The finest result determined by all or certain of the ants delivered in the current cycle is matched with the global finest result and exchanged if it is relevant.
4. An aroma global revising is carried out over the links relevant to the global finest path.

6.2   Clustering Based Trusting Technique

Recent years clustering is widely used in almost all applications. The main objective is how clustering can be used to compute trust in wireless sensor networks. The prime objective is to identify some attacks such as selective forwarding attack, denial of service attack, and sleep dispossession attack where set of nodes remain idle for a long period of time.

The proposed design uses cluster based hierarchical layered architecture that can handle the scalability, flexibility, and energy issues. As it is known that wireless sensor network works in heterogeneous unpredicted environment with low powered sensor devices, the model proves to be very effectual. The model works with four layer hierarchy. It is comprised of one base station on the top, dashed circles depicts

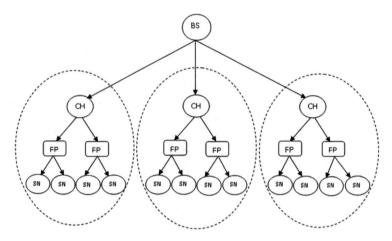

Fig. 5 Cluster based layer hierarchy design for computing trust

clusters, rectangle indicates function point layer, and sensor nodes as leaf node layer. The following Fig. 5 depicts the cluster based architecture for computing trust.

The list of primary building blocks used in the architecture and its characteristics are listed below.

1. Base station layer: This layer consists of base station (BS) which is the target sink for communicating information. All the cluster heads of sensor nodes are connected with it.
2. Cluster head layer: This layer consists of cluster heads (CH) of sensor nodes. The cluster is build based on the sensed values by the sensor nodes. If any two nodes are sensing almost the identical value, they will make a cluster. Cluster can be formed by specifying certain set of wireless limit. If the nodes are within same limit, they will be part of same cluster. The main advantage of making cluster is that it is easy to identify any fluctuation that can occur within a network. In each cluster, cluster head is primary responsible for all action takes place within the cluster.
3. Function point layer: The second most responsible node after the cluster head is the function point (FP) node. These nodes work as an intrusion detection system agent. The decision about malicious nodes is taken primarily based on the information reported by function point. The function point maintains information about the sensor nodes attached to it and maintains the information such as number of packets dropped, number of packets received successfully, number of packets received within time, the period of time the node is idle etc. It can also participate in identifying compromising node which is identified by other function point within same cluster.
4. Sensor node layer: The bottom layer of the model contains all sensor nodes (SN). These are generally responsible for sensing various physical phenomenon

information and for communicating it to the base station by passing through the hops.

The function points have the information of number of nodes connected, number of benign nodes, and number of malicious nodes attached to it. At any point of time, if any sensor node is involved in unusual activity such as drop in packet rates, blocking of wireless frequency, corrupting of packets etc., then the function point of the cluster pass this malicious node information to cluster head. Cluster head in turn communicate this information to other function points in the cluster and base station. Finally, the base station communicates this information to different cluster heads. From cluster head this information pass to respective function points. In the same way the information between any two legitimate sensor nodes is communicated in a cluster based network.

6.3   Key Formation Based Trusting Technique

In wireless sensor network, the initial necessity is to build cryptographic keys for succeeding use. To this end, researchers have projected numerous protocols over many years for wireless networks. The basic characteristics of wireless sensor networks provide earlier protocols which are useless. Various existing wireless sensor devices have restricted to computing power, making asymmetric key cryptographic natives as it is too costly in terms of structure overhead. In addition, the transmission design of wireless sensor networks diverse from classical networks. Wireless sensor nodes may require to framework keys with its neighborhoods and with the information collected from nodes. The easiest way for key formation is the width of wireless network. Unluckily, a single compromise member node in a wireless sensor network will exploit the secret key and thus creates decryption of complete wireless sensor network movement.

One alternative to this concept is to apply a unique shared key to form a fixed link of keys, one pair per transmission nodes. Later delete the width of the keys of wireless sensor network after framing the session keys. However, this alternative of the key-formation procedure do not permit inclusion of new nodes after initial arrangement.

Another alternative is formation of asymmetric key cryptography like Diffie-Hellman key. Its fundamental feature is that a node can frame a secure key with some other node in the wireless sensor network. An alternative way is to pre construct the wireless sensor network with a unique shared symmetric key between every pair of nodes, though it do not range well. In such cases, every node in the wireless sensor network require $(n - 1)$ keys, and $n(n - 1)/2$ keys to maintain the required form of the network.

Loading keys based on trusting base terminal is an alternative choice. In this case, every node is required to share only one key with the base terminal and framing keys with the remaining nodes over the base terminal [37]. This deployment makes the base terminal, one point of breakdown. Since, there is only one base terminal, the network may include resistant container for the base terminal, and relieve the risk of real attack.

Researchers currently proposed arbitrary key pre sharing protocols [38] in which an enormous bunch of symmetric keys is selected and an arbitrary subgroup of the bunch is shared to every wireless sensor member. Two members that want to transmit analyze their bunch to find whether they share unique key; if they do, they apply it to form a session key. Not every pair of members circulates unique key. But if the key formation probability is adequately large, members can still frame keys with adequately various members to acquire a complete connected wireless sensor network. The limitation of this access is that attackers who arrange adequately various members could also rebuild the complete key bunch and break the system.

Alternative design is an arbitrary key pre sharing systems that permits flexible to compromise node, as well as analyzes the tools supporting for asymmetric-key cryptography and more effective asymmetric-key systems like elliptic curve cryptography. Eventually, it is required a secure and effective key sharing system that permits elementary key formation for sensor networks at large scale.

7 Conclusion

The concept of internet of things (IoT) is becoming more pertinent to the realistic world due to the development of mobile devices, embedded and ubiquitous communication technologies, cloud computing, and data analytics. It has the potential to significantly transform several fields such as healthcare, energy management, business, agriculture, manufacturing, transportation etc. It is made up of appliances connected via internet and is capable of gathering information about the environment, communicating the information and to make appropriate action by influencing the environment autonomously with the help of sensing nodes. This leads to wireless sensor networks more visible. Simultaneously, it leads to various risks in the form of attacks. This chapter throughs light on various attacks and its symptoms that can be found in wireless sensor network. It also analyzed the technologies of trust system; emphasize the difference and challenges of trust systems of wireless sensor networks. Also, a brief survey on classical and modern trust systems of wireless sensor networks is presented. Though there exists several models to prevent these attacks, but deep research on these models are welcome in wireless sensor networks. Future research can be planned to design more suitable trust evaluation models for wireless sensor networks that include many advanced types of attacks such as blackmail, buffer over flow attack etc. Besides, theoretical development in order to decrease the energy consumption of trust system can also be planned further.

References

1. Akyildiz, I., Su, W., Sankarasubramaniam, Y., Cayirci, E.: Wireless sensor networks: a survey. Comput. Netw. **38**, 393–422 (2002)
2. Karlof, C., Wagner, D.: Secure routing in wireless sensor networks: attacks and countermeasures. Ad hoc Netw. **1**, 293–315 (2003)
3. Hill, J., Szewczyk, R., Woo, A., Hollar, S., Culler, D., Pister, K.: System architecture directions for networked sensors. In: Proceedings of ACM ASPLOS, pp. 93–104 (2000)
4. Liu, F., Cheng, X., Chen, D.: Insider attacker detection in wireless sensor networks. In: Proceedings of IEEE INFOCOM, pp. 1937–1945 (2007)
5. Li, G., He, J., Fu, Y.: A group-based intrusion detection scheme in wireless sensor networks. In: Proceedings of GPS Workshops IEEE, pp. 286–291 (2008)
6. Xu, W., Ma, K., Trappe, W., Zhang, Y.: Jamming sensor networks: attack and defense strategies. IEEE Netw. **20**, 41–47 (2006)
7. Perrig, A., Stankovic, J., Wagner, D.: Security in wireless sensor network. In: Proceedings of New York, NY, USA, pp. 53–57. ACM Press, New York (2004)
8. Soomro, S.A., Memon, A.G., Baqi, A.: Denial of service attacks in wireless ad-hoc networks. J. Inf. Commun. Technol. **4**, 01–10 (2010)
9. Walters, J.P., Liang, Z., Shi, W., Chaudhary, V.: Wireless sensor network security: a survey. Secur. Distrib Grid Pervasive Comput. **3**, 10–15 (2006)
10. Hu, Y.C., Perrig, A., Johnson, D.B.: Packet leashes: a defense against wormhole attacks in wireless networks. In: INFOCOM, Twenty-Second Annual Joint Conference of the IEEE Computer and Communication Societies, vol. 3, pp. 1976–1986 (2003)
11. Mahajan, V., Natu, M., Sethi, A.: Analysis of wormhole intrusion attacks in manets. In: Proceedings of IEEE Military Communications Conference (MILCOM), pp. 1–7 (2008)
12. Mohammadi, S., Atani, R.E., Jadidoleslamy, H.: A comparison of link layer attacks on wireless sensor networks. J. Inf. Secur. **4**, 69–84 (2011)
13. Lopez, J., Roman, R., Alcaraz, C.: Analysis of security threats, requirements, technologies and standards in wireless sensor networks. Found. Secur. Anal. Des. **57**, 289–338 (2009)
14. Chen, X., Makki, K., Yen, K., Pissinou, N.: Sensor network security: a survey. IEEE Commun. Surv. Tutor. **11**, 52–73 (2009)
15. Law, Y.W., Havinga, P.J.: How to secure a wireless sensor network. In: Proceeding of the International Conference on Intelligent Sensors, Sensor Networks and Information Processing Conference, pp. 89–95 (2005)
16. Kyriazanos, D.M., Prasad, N.R., Patrikakis, C.Z.: A security, privacy and trust architecture for wireless sensor networks. In: Proceedings 50th International Symposium ELMAR, pp. 523–529 (2008)
17. Hoffman, K., Zage, D., Rotaru, C.N.: A survey of attack and defense techniques for reputation systems. ACM Comput. Surv. **42**, 1–31 (2009)
18. Blaze, M., Feigenbaum, J., Lacy, J.: Decentralized trust management. In: Proceeding of the IEEE Symposium on Security and Privacy, pp. 164–173 (1996)
19. Chapin, P., Skalka, C., Wang, X.: Authorization in trust management: features and foundations. ACM Comput. Surv. **40**, 1–48 (2008)
20. Grandison, T., Sloman, M.A.: Survey of trust in internet applications. IEEE Commun. Surv. Tutor. **3**, 2–16 (2000)
21. Momani, M., Challa, S., Alhmouz, R.: Can we trust trusted nodes in wireless sensor networks? Proce. Int. Conf. Comput. Commun. Eng. **3**, 1227–1232 (2008)
22. Lin, C., Varadharajan, V.A.: Hybrid trust model for enhancing security in distributed systems. In: Proceedings of the Second International Conference on Availability, Reliability and Security, pp. 35–42 (2007)
23. Abdul-Rahman, A., Hailes, S.: Supporting trust in virtual communities. In: Proceeding of the 33rd IEEE Hawaii International Conference on System Sciences, pp. 4–7 (2000)
24. Zhou, M., Dresner, M., Windle, R.: Online reputation systems: design and strategic practices. Decis. Support Syst. **44**, 785–797 (2008)

25. Tafreschi, O., Maler, D., Fengel, J., Rebstock, M., Eckert, C.: A reputation system for electronic negotiations. Comput. Stand. Int. **30**, 351–360 (2008)
26. Ziegler, C.-N., Golbeck, J.: Investigating interactions of trust and interest similarity. Decis. Support Syst. **43**, 460–475 (2007)
27. Bo ukerche, A., Ren, Y.: A trust-based security system for ubiquitous and pervasive computing environments. Comput. Commun. **31**, 4343–4351 (2008)
28. Sun, Y.L., Han, Z., Yu, W., Liu, K.J.R.: A trust evaluation framework in distributed networks: vulnerability analysis and defense against attacks. In: IEEE INFOCOM, pp. 1–13 (2006)
29. Qi, J.-J., Li, Z.-Z., Wei, L.: A trust model based on bayesian approach. IEEE INFOCOM Adv. Web Intell. **35**, 374–379 (2005)
30. Josang, A.: A logic for uncertain probabilities. Int. J. Uncertain. Fuzziness Knowl. Based Syst. **9**, 279–311 (2001)
31. King-Casas, B., Tomlin, D., Anen, C., Camerer, C.F., Quartz, S.R., Montague, P.R.: Getting to know you: reputation and trust in a two-person economic exchange. Science **308**, 78–80 (2005)
32. Jaramillo, J., Srikant, R.: Darwin: distributed and adaptive reputation mechanism for wireless ad hoc networks. In: Proceedings of the 13th Annual ACM International Conference on Mobile Computing and Networking, pp. 87–98 (2007)
33. Komathyk, K., Narayanasamy, P.: Trust-based evolutionary game model assisting AODV routing against selfishness. J. Netw. Comput. Appl. **31**, 446–471 (2008)
34. Papaioannou, T., Stamoulis, G.: Achieving honest ratings with reputation-based fines in electronic markets. In: Proceedings of IEEE INFOCOM, pp. 1040–1048 (2008)
35. Caticha, A., Giffin, A.: Updating probabilities. In: The 26th International Workshop on Bayesian Inference and Maximum Entropy Methods, vol. 872, pp. 31–42 (2006)
36. Marmol, F.G., Girao, J., Perez, G.M.: Trims, a privacy-aware trust and reputation model for identity management systems. Comput. Netw. **54**, 2899–2912 (2010)
37. Perrig, A., Szewczyk, R., Wen, V., Culler, D., Tygar, J.: SPINS: security protocols for sensor networks. J. Wirel. Nets **8**, 521–534 (2002)
38. Eschenauer, L., Gligor, V.: A key-management scheme for distributed sensor networks. In: Proceedings of the 9th ACM Conference on Computer and Communication Security, pp. 41–47. ACM Press (2002)

Wireless Sensor Network in Automation and Internet of Things

G.R. Sakthidharan and A. Punitha

Abstract Wireless sensor network in IOT automation and its endowment to enable green computing is of prime concern in recent years. Internet of things (IOT) is the buzz word in today's computing world. It is a collection of devices, connected in the internet that can be controlled, monitored and utilized for a purposeful application. One of the most important elements in the IoT paradigm is wireless sensor networks. The emerging field of wireless sensor networks combines sensing, computation, and communication into a single tiny device. Most often, the connected devices will be obeying the internet protocols and the devices have the capability of sensing the environmental conditions. The devices connected to IOT may use wireless or Radio-frequency identification (RFID), machine to machine (M2M), near frequency communication (NFC), and vehicle to vehicle (V2V) as a connection medium for connecting the host network. The IOT devices sense the environmental conditions using sensor network comprised of sensor nodes. Due to miniature size, minimized data transmission and very low power factor, the sensor devices are attracting and holding a greater space in real-time data computation environments. This makes to concentrate on wireless sensor networks and its benefits towards IOT.

1 Introduction

The Internet of things connects day to day objects like smart phones, Internet, sensors and actuators to the World Wide Web where the devices are smartly linked together, thereby facilitating a new structure of communication between people and things, and between the things itselves. In 2008, the number of connected devices exceeded

G.R. Sakthidharan (✉)
Department of CSE, Gokaraju Rangaraju Institute of Engineering and Technology,
Hyderabad, Telangana, India
e-mail: grsdharan@griet.ac.in

A. Punitha
Department of CSE, Annamalai University, Annamalainagar, Tamil Nadu, India
e-mail: 12charuka17@gmail.com

the connected people [1] and it has been anticipated by Cisco that by 2020 there will be 50 billion connected devices which is seven times the world population. The progress of the IoT will modernize a number of sector, from wireless sensors to nanotechnology. In reality, one of the most vital elements in the IoT paradigm is wireless sensor networks (WSN).

This chapter gives a better understanding of wireless sensor network, components of sensor nodes, real time sensors that are used in current scenario and the software for actual deployment. Also, the role of wireless sensor network in IoT is discussed and finally a brief exploration of WSN in green computing is presented.

2   Wireless Sensor Network

A wireless sensor network (WSN) is a collection of sensor nodes that communicate with each other using radio signals. A sensor node is a small device that contains a sensor, a control circuit and power backup. A sensor node is an intelligent device that has the capability of understanding, analyzing, and transmitting data from one point of route (source) to another point (destination) [2]. It also has the capability of storing and maintaining data via sensor based devices. Its physical appearance makes possible to use in any kind of environment. The need is to design a sensor device which gives a better functionality, better design, better power factor, better interfacing mechanism and importantly, data transfer in less time to the originated destination or sink. When one or more sensors are accommodated in a single system, then the data traffic may increase. But this has to be minimized, such that the time taken should be less, including hopping from node to node, till it reaches the final destination.

The wireless sensor network is not only used to build a versatile and robust application for IOT, also it has a major contribution in green computing. In this digital world, the electronic gadgets get updated within a minimal span of time and this leads to the old gadgets get into scrap. This is impacted in two ways.

1. The electronic gadgets are not utilized for their full life span.
2. The old gadgets that cannot be used further leads to electronic waste.

The increase in the electronic gadgets leads to the consumption of power and processing demand and hence exponentially the electronic waste. Even the processing of a new gadget is underutilized to its capability. If the old gadget components are used for designing small applications, then recycling of electronic gadgets can have a huge consumption by making a green computing revolution.

2.1   Components of WSN Node

The components of WSN node are presented in the Fig. 1. It consists of a sensing unit, a processing unit, a transceiver, and a power unit.

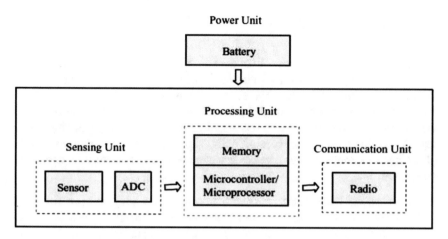

Fig. 1 Components of wireless sensor network node

- *A sensing unit*: The sensing unit contains the vital component of the node and it has the application specific sensors like temperature, pressure, humidity, sound etc., The sensing data may transmit via nearby nodes up to the sink node. If the application is for temperature monitoring, a thermistor will be act as a sensing unit. According to the application, the sensing device will change.
- *A processing unit*: A programmable microcontroller acts as a processing unit. The processing unit controls the entire node operations by sending control commands to the corresponding device. The microcontroller can be selected from range of manufacturers and Atmel, Microchip, Siemens, Texas instruments are some of the market leaders.
- *A transceiver unit*: A transceiver is a device, performing both the operation of a transmitter and a receiver, in a single housing or shares a common circuit.
- *A power unit*: The power unit gives life to the sensor node. This unit accommodates a battery, commonly with a rechargeable option. Advancement in power unit leads to solar charging. A long life power backup with intelligent power aware, enables recharging automatically by solar and also with normal power outlet. This encourages the usage of sensor network in a tremendous application.

2.2 Creating Wireless Sensor Network

When one or more WSN node combines and connects with each other with a sink node/head node, a wireless sensor network is formed. In the WSN, sensors will normally be placed to work for an appropriate application. All the sensors will obey the protocols and hierarchical communication stipulated in the network.

2.3   Classification of WSN

The sensor deployed inside the WSN defines the classification. The classification defines the WSN, whether it uses identical sensor nodes or different functionality sensor nodes with in the network. If the network uses identical sensor nodes for its processing, the computing is simple but if it uses different kinds in the network, then the computing is very complex to receive all the sensor data for processing. In this regard, there are two major types of WSN with respective of using sensor nodes [3] for instance homogeneous wireless sensor networks, and heterogeneous wireless sensor networks.

1. *Homogeneous WSN*: The sensor nodes present in the homogeneous network will have identical hardware, software and network implementation as shown in the Fig. 2. In this type, a single protocol will work and compute for the whole network. The data transmission from node to sink starts from the next near proximity sensor node. This will continue till the data is reaches the designated sink node. So, in each and every data transmission, the node nearby sink node has to work all the time and this leads to more consumption of power resource by a single node. Hence the power backup is drained very rapidly.

   All the nodes that are far away from the sink node have more power backup and enjoy full utilization of power only for its own purpose. But, the node present nearby the sink node has to work for other node to data transfer as well as its own. So, this power utilization is worst compared to the nodes that are present far away from the sink. In a considerable time, the nearby node will wear out in a short span with respective to far away node which rely for a long time. In normal applications, homogeneous network is preferred due to its simplicity to implement and it is easily adaptable to environments. The cost of deployment is also bearable.

2. *Heterogeneous WSN*: In the heterogeneous wireless sensor networks shown in Fig. 3, the sensor nodes present will have unique characteristics. Each sensor node with similar characteristics forms a group. Each group of sensor nodes will have different functionality. The hardware, software and functionality of all groups will be differing. The challenge is to compute the entire datum which is sent from all the groups. So, typically the design of this kind will be complex. But the implementation of all network forms a hybrid network. This kind of network is used to collect different data and have to compute for taking analytical decisions. In this type, every similar node forms a network and will have a head node called cluster head. All the nodes of this network will transmit the data to sink via this cluster head.

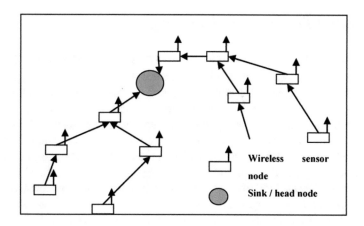

Fig. 2 Homogeneous WSN

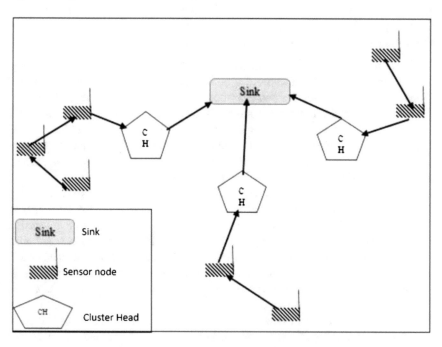

Fig. 3 Heterogeneous WSN

2.4 Sensors for Real Time Processing

Every sensor nodes have a specific defined purpose and functionality. There are varieties of sensor nodes available in the market that can be chosen for a defined purpose. For instance, in home appliance control system, all the controls is based on sensing and analyzing the different parameters. For better understanding, lightings, fans, air conditioners, geyser and pollution control has to be monitored for a better utilization. In this scenario, if a person enters a room, automatically lights have to be switched on. This can be implemented by simply installing a motion sensor. Here, the motion sensor triggers the operation of switching on/off of the electrical appliance. So, the data sent by the motion sensor has to match and bind with the respective room where the person is entering. All the lightings need to be switched on/off according to the predefined conditions. On the other hand, AC can report itself for cleaning its filter and service reminders. For this, AC has to be coupled in a common home network. Even more this application can be smarter; a liked TV program may set for its time and day. This enables efficient utilization of the electrical appliances, conserves electricity.

1. *Sensor node configuration*: Some of the available sensors with their configuration is given in Table 1 to have an idea of internal functionality. The below stated sensor nodes are taken for knowing the type of processor, memory, I/O sensors and the type of radio mechanism used. The configurations mentioned are taken from few real time implementations. Also a specific sensor node can be designed by customizing with basic components and can be used for selecting a better configuration for a specific application. Subsequently, some of the popular sensor nodes used at current scenario is briefly presented.

2. *The ring sensor*: The ring sensor in the Fig. 4 resembles the same as wearable ring. This wearable biosensor consists of low power transceiver and has the ability to accomplish bi-directional communication with a base station. The main function of this sensor is to monitor the physiological status like temperature, blood pressure and heart beat. The patient's current medical conditions are recorded within the biosensor and the measured da-ta are transmitted to a computer through a digital wireless communication link and hence the patient health status is analyzed continuously and remotely.

3. *Smart dust*: It is a tiny device that is used to detect light, magnetism, vibrations, temperature and chemicals. A smart dust is a type of wireless micro electro mechanical sensors (MEMS) in a tiny form factor. They are usually operated on a computer network wirelessly and are distributed over some area to perform sensing through radio-frequency identification. In the Fig. 5, a penny is very large compared to the smart dust.

4. *E - Nose*: A nose-on-chip or e-nose as in Fig. 6 is a single chip that contains both the sensors and the processing components, that recognize or detect the exact components of an odor and analyzes its elemental composition to identify it. Its applications include detection of odors specific to disease for medical diagnosis, and detection of pollutants and gas leakages for environmental protection.

Table 1 Sensor node configuration

Sensor name	Configuration
Tmote	CPU: 8 MHz Texas Instruments MSP430 microcontroller, Memory:10 k RAM and 48 k Flash IO sensors: Integrated Humidity, Temperature, and Light sensors, Radio: 250 kbps 2.4 GHz IEEE 802.15.4 Chipcon Wireless Transceiver
Sensoria WINS 3.0	CPU: Intel PXA255 (scalable from 100 to 400 MHz) Memory: 64 MB SDRAM 32 MB Flash I/O sensors: GPS, USB 2 host ports, 1 device port, RS-232 serial, Audio in/out, GPS, USB, PCMCIA / CardBus, Radio: Dual embedded 802.11b modules
UAMPS MIT	CPU: StrongARM SA-1100, Memory: 16 Mb RAM, 512 KB ROM I/O sensors: Seismic and acoustic sensor, Radio: Interface to the SA-1100 ISM 2.45 GHz with 1 Mbps and range up to 15 m
EYES	CPU: MSP 430F149 5 MHz @ 16 Bit, Memory:60 Kbytes of program memory, 2 Kbytes of data memory, 4 Kbyte EEPROM I/O sensors: UART, AD and I/O, JTAG interface and sensor board with compass, accelerometer, temperature sensor, light sensor, pressure sensor, microphone and push button lines Radio: RFM TR1001 hybrid radio transceiver
IPAQ UCLA	CPU: 206 MHz Intel StrongARM, Memory: 64 MB RAM 32 Flash I/O sensors: Acoustic: built-in microphone and speaker, RS232 serial interface, USB Radio: IEEE 802.11 compliant with 11 Mbps
Imote	CPU: Atmel ATmega 128L, Memory:4 K RAM 128 K Flash I/O sensors: Large expansion connector Radio: 315, 433 or 868/916 MHz Multi Channel transceiver with 38 Kbaud
Mica2	CPU: ARM core 12 MHz, Memory: 64 KB SRAM, 512 KB Flash I/O sensors: USB, UART connector Radio: Bluetooth with the range of 30 m

Fig. 4 The ring sensor

Fig. 5 The smart dust

Fig. 6 E - Nose

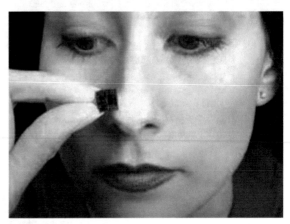

5. *MICA mote*: The core of a mote is a small, low-cost, low-power computer. The computer monitors one or more sensors, including sensors for light, humidity, pressure, temperature, vibra-tion, acceleration etc., The computer connects to the outside world with a radio link. The most common radio links allow a mote to transmit at a distance of 10 to 200 ft. Motes can operate LEDs, the transmission of short radio messages and can communicate with a PC by using serial port. The motes come in two form factors: rectangular and cir-cular. Rectangular mote: Size: 5.7 × 3.18 × 0.64 cm, the size of the mote is sufficient to fit on the top of two AA batteries which give life to the mote. Circular mote: Size: 2.5 × 0.64 cm. This mote is fit on the top of a 3 V button cell as shown in Figs. 7 and 8.

Fig. 7 Rectangular mote

Fig. 8 Circular mote

2.5 Operating System

Operating system (OS) is software which defines the working of hardware components. A survey of operating systems for WSN is found in [4, 5]. After selecting the right configuration, the node should function according to the application and more importantly all the internal and external devices attached with the processor need to be integrated and this is done by the OS. By this the entire design of sensor node gets completion. In this section the design consideration is defined and the OS used in current scenario is briefed.

2.5.1 Design Constraints

The design becomes challenging, when getting to carve an OS for tiny device with tiny memory. There is no constraint in desktop PC OS design, as the memory, processor design and power factor, is available or may include easily in design without any complexity and demand as per aggressive application needs. But in contrast, the design for the sensor node has a lot of constraints. Some important considerations are power factor of processor, memory access time, bandwidth utilization, power management, portability, OS and application compatibility, multitasking and multi processing, network participation, automated routing and scalability.

2.5.2 Tiny OS

Tiny OS is a contribution of open source community based on BSD license [4]. The independent power to have an Opensource OS for developing real time systems is a great boon to the researchers, developers and academicians. The tinyOS supports for low powered wireless devices, such as sensor node, personal area networks and negligible memory devices to fuel the design and development of small or tiny devices. The system, library and applications of tinyOS are written in NesC.

NesC is a new language for programming structured component based system. This language is well suitable for embedded system. The syntax is much like of C. It also exclusively supports the TinyOS concurrency model with capable of structuring, naming and linking the software components into a single powerful embedded system.

2.5.3 SOS

SOS is a dynamic operating system developed for sensor networks [6].The kernel and modules are separated. The SOS supports messaging, dynamic memory allocation, loading and unloading modules. The main advantage of this is, after the deployment of network, the dynamically loaded machine code modules can be replaced. SOS supports run time reconfiguration of embedded software.

2.5.4 Contiki

Contiki is an open source, highly portable, multitasking operating system for memory efficient networked embedded systems and wireless sensor networks. Contiki is designed for microcontrollers with small amounts of memory. A typical Contiki configuration is 2 kilobytes of RAM and 40 kilobytes of ROM. Contiki supports some good sensor list like temperature, door or window opening etc.

2.5.5 MANTIS

The MANTIS is a multithreaded operating system written in C for wireless sensor networking platform. Some key features of MANTIS OS (MOS) are: Developer friendly, C API with Linux and Windows development environment, automatic preemptive time slicing for fast prototyping, diverse platform support including MICA2, MICAz, and TELOS motes, energy efficient scheduler for duty-cycle sleeping of sensor node, small footprint (less than 500 B RAM, 14 KB flash).

2.5.6 Nano-RK

Nano-RK is a kind of real time operating system (RTOS), conceived by Carnegie Mellon University. It is designed to use in sensor network based microcontrollers. The term Nano indicates that the RTOS is small. It is just consuming 2 KB of RAM and only using 18 KB of flash. The "RK" is short for resource kernel. A resource kernel provides reservations on system resources for consumption. The reservation will ensure the all node to meet full battery lifetime and to protecting from overload of entire network due to any node fails. Nano-RK is open source, is written in C and runs on the Atmel-based FireFly sensor networking platform, the MicaZ motes as well as the MSP430 processor.

3 Wireless Sensor Network and IOT

Nowadays, people are living in the world of sensor and WSN is the key enabler for the Internet of Things and it plays a vital role in the internet by collecting the environmental information. At present, the sensors are used in smart phones, cars, emission control, health care, industrial safety, agricultural and horticulture, and the list goes on. Everywhere, some sensors is working for protection, comfort, and safety to help the human being for living a better life. These sensors make the world to sense and analyze by means of triggering some well-defined actions called as automation. To reach its full potential, IOT requires automation.

For instance, a rain sensor in car will automatically trigger the wiper when it sense raining. In this way, the driver of the car need not bother about the switching mechanisms; meanwhile he can more concentrate on driving. In smart phones, the orientation is detected by used gyroscope and accelerometer. An accelerometer measures linear acceleration of movement, while a gyroscope measures the angular rotational velocity. The WSN is not only used for industrial and materialized purposes, also employed in sensing the whole human body activities. The healthcare application uses the most of the WSN and the use of wearable sensors [7] for monitoring various health-related biometric parameters in everyday activities is attracting more interest recently. A market analysis firm, ABI research sensed that the market value of wearable sensors is 400 million market in 2014 and still expecting to go its high fly. Also it forecast that about 80 million for fitness and wellness monitoring. This data is just for health care sector and this shows that the usage and implementation of sensor based applications is in great demand with aggressive growth.

Sensors leads a vital role in taking care of human life. If a disease symptom is identified earlier by means of sensor, it helps to prevent and also to take pre-caution measures by consulting the respective physician. If appropriate sensors are fixed in human body, the respective sensors will analyze, sense and trigger actions when it find any abnormal functionality of human organ.

Figure 9 shows the various sensors for monitoring the human body. ECG sensor node is used to monitor the heart beat and likewise the blood pressure sensor node

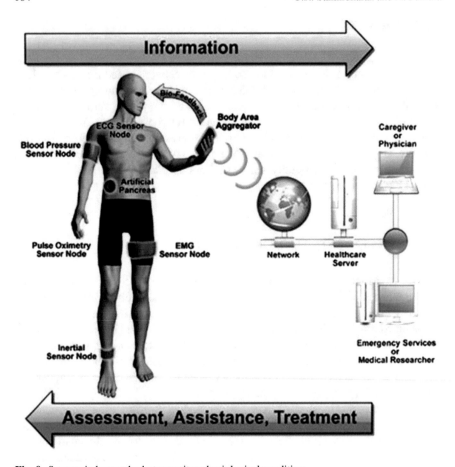

Fig. 9 Sensors in human body to monitor physiological condition

is used to give the details of blood pressure when it is raised above or below the prescribed level. In this, all the sensors form a Personal Area Network (PAN). The PAN is monitored by a central system and in this case, a healthcare server. The data collected in the PAN can be transmitted over internet, which put a first step of Internet of Things, where the resources can manage in remote and trigger a responsive action. By this, IOT come in to existence, where a resource can be monitored, controlled and accessed anywhere, anytime, anybody, in any context and in any network. If sensors can be communicated by any networking or communication mode, then the world of IOT welcomes.

3.1 Integration of WSN with IOT

The challenge in IOT is, to integrate all the WSN elements with other participating devices. As all the elements won't follow a single protocol or communication methodology, there is a need of common platform for all kinds of communications, irrespective of any protocol or standards. This complexity leads the way of finding a better versatile architecture for IOT. The architecture framework is presented in the Fig. 10. It has four layers such as: wireless sensor network, network access, operation and management, and application.

The devices in WSN comprises of RFIDs, Gateways, M2M terminals and sensor node will produce relevant data as per their functionality. The data's were transmitted to the network access layer. In network access layer, various communication modes like mobile, satellite and dedicated line are adopted. This layer plays the vital role in transferring the data from and to the interconnected devices in the IOT framework. The service support system take cares of network and service management. The terminals and the applications can be accessed by operation support system for providing functionalities like user authentication, billing, data analysis and service acceptance. The applications that are linked to IOT framework are present in the application layer. In this way the WSN is integrated with other elements of the network. All the devices forms an IOT. The IOT will be accessed as per the application layer constraints. Even though, all the devices follow some protocols and standards, the IOT gives a fair function and operation.

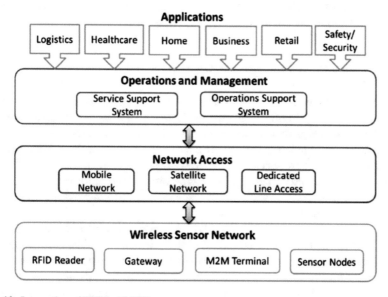

Fig. 10 Integration of WSN with IOT

3.2 Automation of WSN in IOT

Automation or automatic control, is the use of various control systems for operating equipment such as machinery, processes in factories, boilers and heat treating ovens, switching in telephone networks, steering and stabilization of ships, aircraft and other applications with minimal or reduced human intervention. Some processes have been completely automated.

Automation has been achieved by various means including mechanical, hydraulic, pneumatic, electrical, electronic devices and computers, usually in combination. Complicated systems, such as modern factories, airplanes and ships typically use all these combined techniques. In other terminology, automation is a combination of hardware devices and software, that are meant to monitor, control and process a certain task or function according to the predetermined goals. For instance, in automated traffic control system, a central microcontroller collects data's from the sensors node fixed in the respective junctions. All the sensor node installed is connected and controlled by the central microcontroller. The sensor nodes sends the traffic status information to central microcontroller. The central microcontroller will take decision according to the programmed algorithm. The algorithm will work according to the real time data send by the sensor nodes. So, at a particular time if any junction suffers any traffic snares which cannot be controlled by the system, will be communicated to the respective authorities, i.e., the traffic police, to take control of the situation.

There are some important and vital sectors that need to be automated. For better understanding, the sector where automation has a lead role is discussed. These are supply chain management automation, automation of irrigation system, home automation, and traffic automation.

3.2.1 Supply Chain Management Automation

Every nation's economy is depends upon the supply chain management. In India, logistics are not only driving the goods, also it drives the economy. Its contribution is around 6.7% of total gross domestic product (GDP). India has the largest geographical area, so the logistic industry have to ply over to all the parts of the nation. With the increasing importance of services in logistics and supply chain management, there is a need for gathering, processing, and accessing various kinds information throughout the logistical process of a supply chain and such information plays a major role in determining the quality of the product.

The area of coverage and its nature also depends on the better connectivity. For instance, if the logistics is based on hill station bounded area, the cost of fuel and time is crucial. In other words, the kilometers travelled and the fuel expenses won't match, if the hill station kilometers is 25 km, it may take around three hours or more depending on the traffic and also fuel may burn more than 50 km, as the vehicle has to chunk and struggle to move uphill. Inversely, in plain roads, 25 km will take just 30 to 45 min also there may sometime save fuel less than that of the travelling kilometers.

The logistics segment is a peculiar segment, where a system is needed to monitor all the details regarding travel, the goods which it is carrying and the mechanical parts of the vehicle. Here automation plays a vital role. A GPS is needed to track and synchronize the time to reach the destination. Different types are sensors are needed to monitor the mechanics of vehicle, like speedometer linked to GPS. Data has to be computed to get the distance coverage and remaining time to reach the destination. To monitor goods like milk, vegetables etc., a temperature sensor has to be used [8]. If the temperature increases or decreases with reference temperature, the sensor has to trigger an action immediately to correct the lapse, by increasing or decreasing the air conditioner's control. If there is no system of monitoring the temperature of the container, then the consumable goods will be spoiled or contaminated and hence, automation pays profit in this scenario. In this way, WSN can be used for process control and verification in transport and logistics. Internet backbone is necessary to be constructed for the WSN application and when this system can be accessed in any way or any manner by means of respective methodologies using internet, then IOT will be deployed.

3.2.2 Automation of Irrigation System

The water is the prime resource for agriculture and in tropical countries, rain is the only source of water. Water management to crops can be done using irrigation where, irrigation is an artificial application of watering to land or soil [9]. It is used to support the watering to agricultural land, maintenance of landscape etc. Already some established system using WSN was implemented to irrigate and save water at timely intervals. The water consumption varies depending on crop to crop. Appropriately, the water has to be supplied. Else crops may wilt and in extreme case crops fails to yield or die.

An automated irrigation system was developed to optimize the usage of water for agricultural crops. In irrigation system automation, sensors were used to monitor soil moisture and temperature. The sensors were fixed in the root of the crop. The temperature details were computed and connected by WSN and GPRS module can be used to transmit the data to a web. This system is designed as web based and the details of water irrigation plans can be monitored anytime, anywhere irrespective of the geographical location.

3.2.3 Home Automation

Home automation is needed to control, monitor and utilize any kind of home appliances and as technology advances, smart sensor nodes and actuators can be incorporated into appliances, such as vacuum cleaners, micro-wave ovens, and refrigerators. These sensor nodes inside the devices can communicate with each other and with the external network via the Internet. They allow the end users to control home devices remotely and can be used as alarms for disasters at homes. The automation needs to bother about the power consumption and maximum utilization of the home appliances. In earlier systems, wired technology is used to automate the home. But it takes more complex in installation and deployment. Now, wireless technology gives more freedom rather than installing the wires at home. This makes a way to stay connected with any network like web based or PAN or both. Further improvements in the medium of communication technology like RFID, Bluetooth and Wi-Fi were very flexible to use. The automation integrates multi mobile devices based on, networking, distributed and power line communication control to all available home appliances [10]. The user interface may be a mobile device with relevant controlling software, or web-based on desktop or handheld remote, or all of them. More sensors like motion sensor, light dependent resistor sensor, temperature sensor etc., can be used in home automation according to the appliances and application.

3.2.4   Traffic Automation

The traffic in city needs a full control around the clock for effective and consistency flow. As manual traffic monitoring is not an easy task round the clock, automation is the only way [11]. Sensor based automation can be done in a better, more precise and an orderly manner. The traffic inflow and outflow can be monitored by installing two types of sensor networks, the stationary and floating sensor networks. Stationary sensors are wireless sensor devices placed at permanent locations in the ground with traffic sensing capabilities. A mobile or floating sensor is a wireless sensor device attached to a vehicle travelling within the urban sensing field. All the data's from the floating and stationary sensor networks will be analyzed and it is interfaced with a satellite navigation system. The resultant helps in administrating the traffic for smooth flow.

3.3   Verdict of Automation

Automation leads to right resource management and the maximum life time of the equipment, devices and home appliances. The microcontroller technology is changing day to day, to accommodate the minimized or very negligible form factor. 32 bit microcontrollers are dominating the market force with enhanced power factors by consuming very low input voltage. This can be programmed by customizing with respective development environment. The automation converges all kind of available resources and manage the same with a versatile platform, called IOT. Also, the components used in WSN are very compact in size, and most importantly, avoids heat discharging, and helps to reduce carbon foot prints. In this manner, green computing is also achieved. Thus, WSN leads automation, IOT technology and green computing.

4 Green Computing and IoT

Green computing [12] attracted WSN for the environment friendly and energy saving techniques and development processes. Green computing is termed as the practice of using the IT and resources efficiently, reuse, rebuild and hence, promises to less carbon foot with minimal or no bang on the atmosphere. It describes the way of reducing energy consumption [13] of the computers and peripherals. In this modern era, electronic waste is a vital problem that needs to be encountered immediately. Green computing also takes cares of disposing or reusability of the electronic waste. All the WSN will have a certain life span. After expiry of the life span, the WSN can be refurbished for the same or reused for a new application. This makes full utilization of all the components and help to achieve green computing methodology. By reusability, financial investment becomes very negligible.

Every organization uses computers and peripherals for their day to day business processing. All of them is connected in the network by using the network components like routers, switches, hubs etc., all these are electronic components which exhibits more heat on normal operating conditions. In this, for instance the normal operating temperature of Intel core i7 is 50 to 65 C. If an organization has an approximate of 100 systems, then the temperature of the room has to be controlled by cooling the room by using air conditioners. This increases the power utilizations and the electricity bill will be shockingly higher thereby affecting the profit ratio of the organization.

In this case, if sensor for monitoring the temperature [14] is deployed in the computers and in the server room connected by internet, the temperature is totally under the control. In other way, the computers which are not under usage can be shut down and the power supply can be switched off. By this, the wastage of resources and high power consumption scenario can be avoided. Also, solar power grid can be installed to minimize the power consumption. The roof top photovoltaic (PV) panels can be installed for harvesting solar power, where the solar harvesting depends on the direction of the sunlight. So, the solar panels has to be directed towards the sun for better solar power harvesting. In this, a light sensor can be installed to monitor the direction of sun light and the direction of the panel can be rotated by sensing the actual direction. The rotating mechanism should be triggered by moving the solar panel towards the direction of sunlight. The solar power grid can be used for electrical appliances and there is also provision for installing high powered solar grid for high powered appliances like air conditioners, washing machines etc., In industrial areas more solar power grid can be installed and utilized. By installing proper sensors and making them work will improve the utilization of the resources and moreover, some organizations are encouraging their employees to work at home, where the resource utilization is minimized and a better work culture and efficiency is developed. All of the above states that, there are more options for the effective utilization of resources and minimizing the power usage and thereby a green world.

5 Conclusion and Future Challenges

This chapter discussed a review on wireless sensor networks, wearable sensors, few applications of WSN, and integration with IOT. In this era of computing and mobility, the IOT has to make a better convergence of WSN. The real prospective of sensor networks can be fully unleashed if they are connected to the internet. The future of IOT depends on the WSN and its wrathful utilization. One of the challenges in this area is to promise the subsistence of certain security issues in the integration of WSN with the IoT, such as integration of security mechanisms and services, user's acceptance and management of data privacy. As innovative application areas for WSNs in the field of digital image processing are emerging, the WSN operating system needs to offer a rich set of basic image and signal processing APIs to facilitate the responsibility of the application program developer. Also, many investigations can be focused on integrating the various sensor networks to the Cloud for many real-time applications.

References

1. Swan, M.: Sensor mania! the internet of things, wearable computing, objective metrics, and the quantified self 2.0: a survey. J. Sens. Actuator Netw. **1**(4), 217–253 (2012)
2. Sakthidharan, G.R., Chitra, S.: A survey on wireless sensor network: an application perspective. In: Proceedings of IEEE International Conference on Computer Communication and Informatics, pp. 1–5 (2012)
3. Uplap, P., Sharma, P.: Review of heterogeneous/homogeneous wireless sensor networks and intrusion detection system techniques. In: Proceedings of Fifth International Conference on Recent Trends in Information, Telecommunication and Computing, pp. 22–29 (2014)
4. Adi Mallikarjuna Reddy, V., Phani Kumar, A.V.U., Janakiram, D., Ashok Kumar, G.: Operating systems for wireless sensor networks: A survey technical report. Technical Report, pp. 1–30 (2007)
5. Farooq, M.O., Kunz, T.: Operating systems for wireless sensor networks: a survey. Future Gener. Sens. **11**(6), 5900–5930 (2011)
6. Chieh, H., Chih, K., Ram, S., Roy, K., Eddie, S., Mani, K.: A dynamic operating system for sensor networks. In: Proceedings of the Third ACM International Conference on Mobile Systems, Applications, and Services, pp. 34–39 (2005)
7. Jovanov, E., Price, J., Raskovic, D., Kavi, K., Martin, T., Adhami, R.: Wireless personal area networks in telemedical environment. In: Proceedings of the Third International Conference on Information technology in Biomedicine, pp. 272–275 (2000)
8. Evers, L., Havinga, P.J.M., Kuper, J., Lijding, M.E.M., Meratnia, N.: Sensor Scheme: supply chain management automation using Wireless Sensor Networks. In: Proceedings of IEEE Conference on Emerging Technologies and Factory Automation, pp. 448–455 (2007)
9. Gutierrez, J., Villa-Medina, J.F., Nieto-Garibay, A., Porta-Gándara, M.A.: Automated irrigation system using a wireless sensor network and GPRS module. IEEE Trans. Instrum. Meas. **63**(1), 166–176 (2014)
10. Vinay, S., Kusuma, T.: Home automation using internet of things. Int. Res. J. Eng. Technol. **2**(3), 1965–1970 (2015)
11. Skordylis, A.: Information propagation in traffic monitoring sensor networks. Thesis Report, pp. 1–134 (2007)

12. Pande, M., Choudhari, N.K., Pathak, S., Mukhopadhyay, D.: H2E2: a hybrid, hexagonal and energy efficient WSN green platform for precision agriculture. In: Proceedings of 12th International Conference on Hybrid Intelligent Systems (HIS), pp. 67–74 (2012)
13. Chithra, J., Vijay, A., Vieira, D.: A study of green computing techniques. Int. J. Comput. Sci. Inf. Technol. Res. **2**(2), 238–242 (2014)
14. Gubbi, J., Buyya, R., Marusic, S., Palaniswami, M.: Internet of things: a vision, architectural elements, and future directions. Future Gener. Comput. Syst. **29**(7), 1645–1660 (2013)

Challenges of Distributed Storage Systems in Internet of Things

Madhvaraj M. Shetty and D.H. Manjaiah

Abstract Internet of things presents a general concept of the strength of different network devices to sense and collect important information from the world around it, and then it share this information across the internet where it accessed, processed and utilized for various other interesting purposes. This concept of internet of things will be an integrated part of the future internet. But in order to gain more benefits, it is essential to store heterogeneous data captured by various devices for further analysis. But it requires vast amount of storage space which cannot be handled using traditional centralized storage system. So it is achieved using the concept of distributed storage systems and the growth of these storage devices impose many challenges related to topologies and interconnection protocols, data consistency, error handling, security and privacy, etc. Ensuring security and privacy to the users is a major challenge considering the sensitive, personal or confidential nature of the data stored in the system. This chapter, discusses various issues related to fusion, privacy, security and trust in distributed storage systems.

1 Introduction

Internet of Things (IoT) is a way of describing a network of smart devices that connect with each other through the internet to make environment more intelligent. It was coined more than 10 years ago by the industry researchers but has emerged recently into mainstream public view [1, 2]. IoT technology includes sensors, circuits, communications, embedded systems, intelligent interfaces, data management, energy management, distributed processing, data fusion, knowledge management, real-time processing, system design, and sophisticated software techniques that provides great convenience for human beings for their life style [3–5]. It presents a general concept

M.M. Shetty (✉) · D.H. Manjaiah
Department of Computer Science, Mangalore University, Mangalore, India
e-mail: madhvarajj@gmail.com

D.H. Manjaiah
e-mail: manju@mangaloreuniversity.ac.in

© Springer International Publishing AG 2017
D.P. Acharjya and M. Kalaiselvi Geetha (eds.), *Internet of Things:
Novel Advances and Envisioned Applications*, Studies in Big Data 25,
DOI 10.1007/978-3-319-53472-5_9

of the strength of network devices to sense and collect important information from the world around us, and then it share this information across the internet where it accessed, processed and utilized for various other interesting purposes. Internet of things concept will be an integrated part of the future internet and could be defined as a global dynamic network infrastructure having self configuring capabilities based on some standard protocols for communication where virtual and physical things have their identities, attributes and virtual personalities and provides intelligent interfaces, and these are seamlessly integrated into the information network present today [1, 2]. But in order to gain more benefits from IoT, it is necessary to store the data which are generated and captured by these devices. Thus, to achieve storage for Internet of Things, especially the heterogeneous data (structured data, semi-structured data and unstructured data), it requires vast amount of storage space. But traditional centralized storage system is not effective with this size of data, because it is not possible to store everything on single large disk. If this disk fails, then everything will be lost. To overcome all of these issues, the distributed storage came into existence.

The forthcoming part of this chapter is divided into four main sections. The second section provides introduction to the distributed storage systems with the explanation of the types, followed by different architectures available today. The third section introduces challenges related to data fusion. The fourth section discusses issues involving privacy and security. The last section discusses the necessity of trust management in distributed storage systems, followed by conclusion of the chapter.

2 Distributed Storage Systems

Distribution is the key for handling huge data sets and it is essential to bring scalability in today's world, i.e., maintaining stable performance for growing data collections by adding new storage resources to the existing system. Distributed data storage has become more popular for efficient data management and significantly evolved due to cloud computing upsurge in the past several years, as the underlying core infrastructure for the cloud computing is distributed storage systems like the Hadoop Distributed File System (HDFS) which are the foundation of all types of cloud services. Basically, distributed storage systems makes use of the same main elements as centralized systems does, but in a distributed manner (i.e., data units and its meta data); and therefore different data units are scattered over the networked system: files, objects, blocks or data chunks. The main purpose of distributed networked storage systems is to store the data consistently over a long period of time by using storage servers that are distributed [6, 7]. But Long term reliability requires some sort of redundancy, a straightforward solution of simple replication is used; however, the cost of storage system is high.

Nowadays, distributed storage systems are very popular and extensively used throughout the world in many applications. The growth of these storage devices impose many challenges related to topologies and interconnection protocols, data consistency, error handling, security etc. Also this distribution brings a number of technical problems that makes the design and development of distributed storage, indexing, searching and computing a delicate issue. Often, this distributed storage systems may contains sensitive data, and thus the security methods are crucial. Since the stored data has to be shared and replicated in a secure way, such architecture poses number of security challenges. This security challenge became more complicated with the widespread adoption of distributed storage systems and it is absolutely essential to understand these challenges. In addition, it is also extremely challenging, due to the reason that malicious adversaries can reside anywhere and attack this distributed storage locations because systems are connected over the internet.

The main objective of distributing storage across multiple nodes is to make data available closer to the users and applications and to protect the stored data in case of failure of disks through redundant storage in multiple devices across the system [8]. Popularity of distributed storage systems has been increased due to the rapid growth of bandwidth, storage volume and computation resources along with the reduction in the cost of storage devices.

2.1   Types of Distributed Storage Systems

There are mainly four different types of distributed storage systems: Network Attached Storage (NAS), Server Attached Redundant Array of Independent Disks (RAID), Storage Area Network (SAN) [9] and Centralized RAID as shown in Fig. 1 [8]. Among the four distributed storage system listed here, NAS and SAN are the most popular systems where NAS uses TCP/IP protocol and SAN uses SCSI protocol for communication. While NAS can be implemented on any physical network which supports TCP/IP such as Ethernet, fiber distributed data interface (FDDI), or asynchronous transfer mode (ATM). Storage area network can be implemented only on networks with fiber channels. Compared to SAN, NAS has lower performance, because of the fact that SCSI is faster than TCP/IP networks and TCP has higher overhead.

Fig. 1 Types of distributed storage systems

2.2 Distributed Storage Architectures

There are multiple implementations of distributed storage systems. The most important architectures among them are: Google File System, Lustre, General Parallel File System, Andrew File System and Ceph.

2.2.1 Andrew File System

Andrew file system (AFS) is a pioneer and one of the top most representative of networked attached storage that uses files (i.e., data units) to distribute data over multiple nodes in network. It is a location independent file system, makes easy for users to work together on the same data without knowing the location where the file actually stored. AFS organize data units in cells and volumes. Cells are just logical divisions of file space and volume can be seen as a partition managed by a server. Individual files are organized hierarchically in sub-trees which are grouped into cells. In AFS, redundancy is achieved by using a special type of volume or partition which is named as read-only copy, used to acquire high availability and workload balance.

There are three major implementations of AFS namely: OpenAFS, Transarc (IBM), Arla and fourth implementation exists in the Linux kernel source code committed by Red Hat. Advantages of AFS is that, it performs well in small as well as very large installations, provides exceptional performance and provides high security control based on Kerberos.

2.2.2 Google File System

Google file system (GFS) [10] was built to organize and manipulate huge amount of data capitalizing strength of reliable, expensive servers while compensating cheap hardware weakness. It is designed to provide reliable access to data using large clusters of commodity hardware efficiently. It is not an open source system, but serves as model for other systems for example HDFS: file system used by hadoop [11]. In essence, GFS architecture is composed by one active master GFS server and few clones such as for fail over scenarios, to manage the file systems meta data, and hundreds of GFS slave servers to store chunks of data in a distributed manner. The system provides series of mechanisms to support huge number of clients requests that can access shared file system in parallel consistently. Benefits of GFS is its simple design, fault tolerance, and effient, reliable access to data using commodity hardware. Limitations is that it is only viable in a specific environment and has limited security.

2.2.3 General Parallel File System

General parallel file system (GPFS) [12] is a parallel, scattered file system solution from IBM. In contrast with GFS, GPFS supports RAID systems over fiber channel SAN. Also it has the ability to use LAN systems through NSD (Network Shared Disk) which enables block access to the data. The choice between these two modes depends on the system purpose. Accessing data using SAN is much faster, but it requires expensive network equipments, while LAN access is less expensive, but much slower than SAN. GPFS supports deployments of thousands of storage nodes which spreads files across them for achieving a high availability. It also supports replication, logging and recovery capabilities, thus is a fault tolerant system. GFS has no data placement policy defined, while GPFS distributes data by means of two modes of policies for instance file placement policy and file management policy. File placement policy distribute files to specific set of disks (disks pool) and file management policy move or replicate files across system disks. The file system name space can be split into small groups by means of file sets achieving management at a smaller granularity. Another unique feature of GPFS is that the meta data management which spread across all GPFS clients, thus a single point of failure is avoided. It is also highly configurable and supports three deployment types such as share-disk, networked I/O, and multi cluster.

Advantages of GPFS are massive namespace support, seamless capacity expansion and performance scaling, reliability, parallelism, consistency, high availability, and flexible architecture. Virtualizing file storage space, and allowing multiple applications to share common pools of storage provides flexibility to administer the infrastructure without disrupting applications, thereby improving energy efficiency, and reducing management overhead.

2.2.4 Lustre File System

Lustre [13] is open source software licensed under the GPL. It is an advanced storage architecture that provides scalability, significant performance and flexibility to computing clusters and networks. Lustre uses object storage model for I/O and storage management. It uses sophisticated replication, cutting-edge fail over and recovery techniques to eliminate downtime and to maximize file system availability, thereby maximizing productivity and performance. It is composed by three main entities: object storage servers (used to manage objects stored on targets), meta data server (for name spaces management) and management server (used to manage all lustre file systems). In contrast with GFS and GPFS, it splits files into objects. Each object storage server can have a fail over configuration achieved by using more targets to achieve high availability and fault tolerance of meta data and data stored. Both metadata servers (MDS) and management servers (MGS) use similar mechanism. Lustre supports SAN based storage using commodity storage. The data objects are managed by keeping a map similar to the inode map in the Berkeley fast file system (BFFS). But this can be considered as a drawback since it puts more pressure on the storage servers. Lustre's key attributes are its performance, scalability, manageability, flexibility, reliability, availability, and serviceability. The open scalable file systems (OpenSFS) and European open file system group (EOFS) are two groups that support vendor neutral development and promotion of lustre file system.

2.2.5 Ceph File System

Ceph is a new open source distributed file system available in Linux kernel. It brought together many advantages from previous implementations and it distributes everything (name spaces, data storages, logging, monitoring) making it very scalable, but it is very complex as well, and it is being considered as the new dream distributed file system. Ceph addresses three critical challenges of storage systems: performance, scalability and reliability by occupying a unique point in the design space. Its architecture of meta data management addresses one of the most difficult problems in highly scalable storage systems. This provide a single uniform directory hierarchy which can obey POSIX semantics with performance that scales with the large number of meta data servers. Ceph's dynamic sub tree partitioning is a uniquely scalable approach which offers efficiency and ability to adapt to varying workloads. Additionally, it automatically balances the file system to deliver maximum performance and also file system provides numerous benefits. It provides stronger data safety for mission-critical applications and provides virtually unlimited storage to file systems.

It is necessary to have exceptional technologies to process large amount of data present today. Hadoop is one of the most widely used technologies which supports programming model to handle this large amount of data using Map Reduce in distributed manner. Hadoop uses new file system called hadoop distributed file system (HDFS), which is scalable, distributed and portable file system written in Java. It is a implementation of Google file system (GFS) where all the resources stored in Hadoop cluster without any schema representation. Each hadoop cluster contains one master node and several slave nodes. The master node consists of name node, data node, task tracker, and job tracker where slave node acts as both a data node and task tracker which holds computation and the data. The job tracker manages all the job scheduling.

3   Data Fusion in Distributed Storage Systems

Today Information systems collects data, processes it, provides output, and stores these results for future use. In traditional systems, it gets input only from a variety of standard input and use a variety of standard output devices. But in recent systems, it is not same, if we consider today's sensor-based smart world systems, it uses more specialized input devices called sensors and specialized output devices called effectors, which can cause some action to be taken later. For example, sensors attached to the aircraft may detect a position, speed, and orientation; the system would determine any deviation from the desired path; and output signals would be sent to the appropriate effectors to take further action. Also sensors may provide a single binary reading (door is open or closed), a single continuous reading (pressure, temperature, or acceleration), or a set of related readings (multiple readings at a point from a specific location, or an image). So data fusion is essential for these type of system where the multiple types data is distributed over the network, it integrates data from multiple sources to provide better analysis and decision making. Data fusion can be defined as the process of merging of multiple heterogeneous data sets into one homogeneous representation to be better processed for data mining and management, and it is used in a number of technical domains such as sensor networks, video and image processing, intelligent systems, robotics, and other domains.

Data integration is differentiated from data fusion, in integration, more broadly data sets combined and retains the larger set of information while in data fusion, there is usually a replacement or reduction technique and it is facilitated by data interoperability (ability for two systems to communicate and exchange their data). Data fusion and data integration is together forms new set of techniques used for business intelligence, such as integrating online storage of sales information to create more completed pictures for the customers. Data fusion systems face two folds of challenges, first, data from multiple sources are often heterogeneous, it can exist at the schema level or instance level and second, different sources can provide conflicting data, because of erroneous data, incomplete data and out-of-date data. But advances in data fusion can also benefit in many other important areas such as health care (for CAT scans and other types of medical imaging), manufacturing (for robotic and adaptive process control), environment (for the identification and tracking of pollutants), transportation (for intelligent traffic control) and information analysis for intelligence agencies, etc.

4 Privacy and Security

A single isolated storage node is less vulnerable to be victim of malicious activities than the storage which are distributed over multiple nodes. It is desired that data stored in the system must remain private even if all storage servers are compromised. The major challenge of designing these networked distributed storage systems is to guarantee privacy preservation of users while maintaining its distributed structure. In current context, with more and more users, devices and applications are joining the internet, the world has reached the next level where the global network is not just used for disseminating information from a server. Today applications are being run from the cloud and vast amount of data uploaded and saved on the cloud using distributed storage system. With all this technological improvement, malicious attacks with the intention to damage or steal data have also become increasingly common. But level of security has not grown as fast as the systems and applications has grown.

In modern world, traditional approach towards solving some issues security fails due to the ever increasing complexity of security breaches. If a computer is connected to a network, it can get infected by some malicious code easily. It can also cause serious damage to data on memory or might monitor the system user, even when hacker has no physical access to machine. Often in today's world there is increase in sensitive data, such as customer records, health care records or financial data. Protecting such data while they are in transit as well as at rest is crucial. During its life-cycle, the data travels through various networks and storage systems, from various users and ends up in off line or online data archives. Hence, there exists large number of potential attack points. So, the data needs to reliably stored and protected at every stage of its life-cycle. And the data has to be replicated, shared with multiple users and kept online in order to qualify various performance, recovery and availability requirements. As a result these data stored in storage systems are becoming more vulnerable to security breaches, which can result in potential damage and losses. There are more number of threats available in distributed storage systems due to the reason that these systems are inherently more open and exposed to attacks than centralized storage systems, the main reason is that, being the unsecure network of loosely connected nodes. Several active researches are going on in the area of threat modeling and developing security model for protecting distributed storage systems since it contains sensitive data of users which are private. As the adoption of Internet of Things becomes universal, data that is collected and stored becomes huge. The fact that in the IoT, a lot of data flows autonomously without human knowledge makes it very important to define authorization protocols in place to avoid damage and misuse of data.

As the Internet of things becomes a key for the future internet and a critical national/international infrastructure, the need of implementing adequate security for the IoT infrastructure becomes significant and more important. Large-scale applications and services based on the IoT are increasingly vulnerable due to different attack or information theft. Advances are required in several domains to make the IoT secure from those with malicious intent, including.

- At present DoS/DDOS attacks are well understood for the current internet technologies, but the Internet of Things is also vulnerable to such attacks and will require specific mechanisms and techniques to ensure that energy, transport, city infrastructures cannot be disabled.
- General attack detection and recovery - specific threats, such as malicious code, compromised nodes, hacking attacks must be detected and recovered.
- Cyber situation awareness tools/techniques need to be implemented, to enable IoT based infrastructures to be monitored continuously.

Active tap, passive tap, denial of service, faking, replay, traffic analysis. Table 1 shows summary of different sources of security threats and its respective measures can be taken in distributed storage systems, out of these measures authentication, authorization and cryptography acts as major challenging issues in the field of IoT.

Authentication is a fundamental service for security. AFS uses a slight variation of Kerberos (v4) protocol for mutual client server authentication [14]. Lustre [13] is more flexible and supports GSS-API back end, like LIPKEY, Kerberos, and OPEN. PAGs (Process authentication groups) are used to organize processes that are linked

Table 1 Security threats source and it measures in distributed storage systems

Security threats	Corresponding security measures
Storage devices	Redundancy, recovery, backup
Storage systems	Authentications, authorization, access control, virus protection, intrusion detection/protection, firewall
Storage network	Access control, cryptography, authentication and authorization, intrusion detection/protection
Storage application	Authentication and authorization, cryptography, access control, immutable, logging, audit and accounting, tamper-proof

to an authentication event. This can provide a more fine-grained control than normal uid-based mechanism. Lustre and AFS use PAGs, while NFS does not support PAGs, being vulnerable to root set uid attacks. Authorization or access control is performed by comparing the identity of the client with the ACL attached to the object/resource being accessed. Although POSIX.1e access control lists have removed some of the limitations of the initial version of POSIX model, being supported on ext2, NFS etc., it still has not able to meet today's requirements. Some of the DFSs implement their own non-standard access control models, for example, AFS uses group inheritance for easier administration. Positive and negative access control lists are associated with directories not files. Venus clients emulate POSIX semantics for files, which are ignored by other servers. Existing authorization mechanisms fail to provide robust and powerful tools for handling security issues at the scale necessary for today's world of Internet of Things. Securing data stored on the disk should be consider seriously than securing data on the wire due to several factors: securing data on the disk at user endpoints provides confidentiality for both data at rest and in transit, allowing outsourcing of servers for storage services to untrusted servers (public cloud storage), limits the damage done by theft, offers better performance when using optimizations such as automatic revocation of authorizations, since compute-intensive operations are performed at the client side. AFS provides data confidentiality on the wire (i.e., encrypted communication between a client and a server - ex: authentication) but data on the disk is not secured. But in more recent distributed file systems for example Lustre are able to make use of securing data on the disk as well. Thus, the effort shifts to designing efficient key management (distribution, revocation) schemes, access control models having the right granularity, searching and sharing encrypted data.

Also the most important challenge in convincing users to adopt upcoming technologies such as IoT is the protection of their data and preserving privacy, since much of the information in IoT system are personal data which are sensitive, so there is a requirement to support anonymity and restrictive handling of users personal information. There are a number of areas of IoT security where advances are required, Mainly:

- Cryptographic techniques that can enable encryption on the data to be stored processed and shared, without the information being accessible to other third parties. Techniques such as homomorphic and searchable encryption mechanism are potential applicant for developing such approaches.

- Techniques to preserve privacy of the users by designing concepts, including data identification, minimization, authentication and anonymity.
- Self configuring and fine-grain access control mechanisms should be developed for IoT.

5   Trust in Distributed Storage Systems

As Internet of Things applications and services scale over several administrative fields and involves in multiple ownership regimes, there is a increasing need for an trust framework that can enable the user who use the system to have confidence that the services and information being exchanged can absolutely relied upon. Trust is an important concept in distributed storage environments where it plays a critical role to enhance and ensure system security through trust management so that effective and valid information services can be provided to the end users. Although different trust models have been proposed for distributed storage systems and very less research has been conducted to describe trust in formal and precise way. Also formal, clear trust definition is very critical, it can help us to interpret the meaning of trust without ambiguity and implement it on distributed storage and provides good compatibility for extensive collaboration among various computing systems.

In our society, the capabilities of an individual (or an organization) are so limited that they must rely upon and cooperate with other entities in order to achieve various targets of daily life and businesses. This interdependence on each other makes trust come up as one basic social requirement, by implementing this effectively which can enable us to collaborate with others without fear, and lets us to use trust as a key component for successful conflict resolution [15]. Trust Management concept first introduced by Blaze et al. [16], it is a unied approach to specify and interpret security policies, relationships, credentials and that allows direct authorization for critical security actions. Credentials describes specic delegations of trust among public/private keys, unlike traditional certificates, which binds keys to names, trust-management credentials binds keys directly to authorization process to perform specic tasks. Trust-management systems support delegation and policy specification and refinement at the different layers of a policy hierarchy, thus solving a large degree of scalability and consistency problems inherent to the traditional access control lists. Furthermore, trust-management systems design is extensible; it can guide us to define policies for different types of applications.

6 Conclusion

The Internet of Things (IoT) is the interconnected set of intelligent objects and devices designed specifically to enable the virtualization of the physical world. It presents a general concept of the strength of different network devices to sense and collect important information from the world around us, and then it share this information across the internet where it accessed, processed and utilized for various other interesting purposes. Nearly every type of business today can employ IoT systems to create vast amount of cost savings and efficiencies. It will revolutionize the way at which people work, entertain, monitors health, interact with others, manages homes and things around. This will be an integrated part of the future internet, but in order to gain more benefits from IoT, we need to store the data, especially the heterogeneous data which are captured by these devices, it requires vast amount of storage space which cannot be handled using traditional centralized storage system, so it is achieved using the concept of distribution. Distributed data storage has become more popular for efficient data management, significantly evolved due to technologies like cloud computing upsurge in the past several years. The growth of these storage devices impose many challenges related to topologies and interconnection protocols, data consistency, error handling, data fusion, security and privacy, so on. Ensuring privacy and security has become major challenge in the context of distributed storage systems, because it contains sensitive data which are private. It became more complicated with the widespread adoption and it is absolutely essential to understand these challenges for today's world. Advances are required in several domains to make IoT works efficiently and to secure it from malicious intent.

References

1. Evangelos, A., Kosmatos, A., Nikolaos, D.T., Anthony, C.B.: Integrating RFIDs and smart objects into a unified internet of things architecture. Adv. Internet Things **13**(2), 45–55 (2011)
2. Atzori, K., Luigi, D., Iera, A., Morabito, G.: The internet of things: a survey. Comput. Netw. **54**(5), 2787–2805 (2010)
3. Li, H., Lin, X., Yang, H., Liang, X., Lu, R., Shen, X.: EPPDR: an efficient privacy-preserving demand response scheme with adaptive key evolution in smart grid. IEEE Trans. Parallel Distrib. Syst. **25**(8), 2053–2064 (2014)
4. Dong, M., Ota, K., Li, H., Du, S., Zhu, H., Guo, S.: RENDEZVOUS: towards fast event detecting in wireless sensor and actor networks. Wirel. Sens. Netw. Secur. **23**(5), 230–245 (2012)
5. Li, H., Lin, X., Yang, H., Liang, X., Lu, R., Shen, X.: An efficient merkle-tree-based authentication scheme for smart grid. IEEE Syst. J. **8**(2), 655–663 (2014)
6. Rhea, F., Sean, H.: Maintenance-free global data storage. IEEE J. Internet Comput. **5**, 40–49 (2001)
7. Dabek, F., et al.: Wide-area cooperative storage with CFS. J. ACM SIGOPS Oper. Syst. Rev. **35**(5), 202–215 (2001)
8. Hu, A., Yuchong, D., Xi, W.: Cooperative recovery of distributed storage systems from multiple losses with network coding. IEEE J. Sel. Areas Commun. **28**(2), 268–276 (2010)

9. Yu, W., Gao, X., Li, W.X.: A new network storage architecture based on NAS and SAN. In: Proceedings of 10th International Conference on Control, Automation, Robotics and Vision, pp. 345–349 (2008)
10. Ghemawat, H., Sanjay, K., Gobioff, H., Leung, S.T.: The google file system. J. ACM SIGOPS Oper. Syst. Rev. **37**(5), 345–356 (2003)
11. Shvachko, K., Konstantin, L., Ramesh, H.: The hadoop distributed file system. In: Proceedings of IEEE 26th Symposium on Mass Storage Systems and Technologies, pp. 234–240 (2010)
12. Schmuck, D., Frank, B., Haskin, R.L.: GPFS: a shared-disk file system for large computing clusters. J. Comput. **2**(3), 112–121 (2002)
13. Braam, G., Peter, J.: The lustre storage architecture. J. Sci. Environ. **11**(4), 67–77 (2004)
14. Satyanarayanan, K., Mahadev, F.: Integrating security in a large distributed system. ACM Trans. Comput. Syst. **7**(3), 247–280 (1989)
15. Lewicki, H., Roy, J., Tomlinson, E.C.: Trust and trust building. J. Beyond intractability **5**(3), 305–315 (2003)
16. Blaze, F., Matt, K., Keromytis, A.D.: The key note trust-management system version 2. J. Sci. Technol. **12**(2), 56–67 (1999)

Part III
Challenges, Issues and Healthtcare Applications of Internet of Things

Internet of Things Based Intelligent Elderly Care System

J. Arunnehru and M. Kalaiselvi Geetha

Abstract The World Health Organization (WHO) reports that most common cause of injuries to elderly people increases every year due to fall events. Human fall events are one of the most important health problem among the elderly people whos aged 65 and above, which could probably result in a significant barrier to their independent living. This chapter presents a method for fall activity detection based on Motion Projection Profile (MPP) features extracted from temporal difference image to represent a various levels of a person's posture. Falls are detected by analyzing the projection profile features consist of the measure of motion pixel of each row, column, left diagonal and right diagonal of the temporal difference image and gives adequate information to recognize the instantaneous posture of the person. The experiments are carried out using publicly available fall detection dataset and the extracted MPP feature set are modeled by the various machine learning methods like Support Vector Machine (SVM) with polynomial kernel, SVM with Radial Basis Function (RBF) kernel, K-Nearest Neighbor (KNN) and the Decision tree (J48) algorithm are used to classify the fall activies. Experimental results show that SVM with RBF kernel is an efficient to recognize the fall activity with an overall recognition accuracy of 89.55% on the fall detection dataset, which outperforms other machine learning methods.

1 Introduction

Falls is most common among elderly people, disable persons and patients which cause severe domestic accidents such as head injuries and bone fractures that may even result in human fatality. The World Health Organization (WHO) says that an injury due to falls is the major cause of casualty for people over 75. WHO global report on falls prevention in older age, 2007 defines falls as "inadvertently coming to

J. Arunnehru (✉) · M. Kalaiselvi Geetha
Department of CSE, Annamalai University, Chidambaram, India
e-mail: arunnehru.aucse@gmail.com

M. Kalaiselvi Geetha
e-mail: geesiv@gmail.com

© Springer International Publishing AG 2017
D.P. Acharjya and M. Kalaiselvi Geetha (eds.), *Internet of Things: Novel Advances and Envisioned Applications*, Studies in Big Data 25,
DOI 10.1007/978-3-319-53472-5_10

207

rest on the ground, floor or other lower level, excluding intentional change in position to rest in furniture, wall or other objects". Older community express fall as a failure in steadiness/stability. While the health care expert refer these incidents as injury causing events leading to ill health. Besides, fall may lead to loss of independence and sovereignty, mystification, immobilization and despair which will direct to added constraint in the everyday activities. Further, most of the elderly people may not be able to get up without others' assistance and any ensuing long lie after an injury may lead to dehydration, bronchopneumonia, hypothermia and pressure sores [1, 2], and requires immediate medical assistance.

Fall detection techniques may be classified under three main categories [3]: vision-based, environmental devices and wearable sensors. Use of wearable devices [4–7] mounted on clothes or body with embedded sensors, such as accelerometers, nanowatches and gyroscopes, to detect the location and motion of a particular person is commonly seen in the literature. Elderly persons are already sick and making them to wear sensors is an added strain to them. Besides, asking them to use devices for calling assistance after a fall is impractical, since they may be unconscious after such incident. Since these devices are inconvenient for patients and elderly people, studies were carried with triaxial accelerometer in smart phones [8]. But usage of smart phones is also not common among elderly which creates another challenge. Hence, there is a requirement for a self-directed fall detection technique that triggers an alarm without human intervention, so that the injured person could get immediate medical attention. Hence it is necessary to develop an autonomous fall detection technique that overcomes all these limitations.

Further, a fall detection approach must be able to discriminate between a fall and the daily activity performed by a person. This is not an easy task, since certain movements such as going from standing posture to lying down posture have strong similarity to a fall event. Hence, the problem of fall detection is viewed as activity recognition in this proposed work. This chapter analyses the fall detection problem as the vision based activity recognition. In vision-based activity recognition, the procedure is carried at four steps viz. human detection, human tracking, activity recognition and then a complex activity assessment to evaluate whether the event is a fall or not.

Survey on recent techniques in human fall detection can be found in [9]. Willems et al. [10] presents an outline of various approaches for fall detection using 2D cameras. Initially, they perform background subtraction and used the motion information of the person for fall identification. Audio based fall detection approach is proposed by Planinc and Kampel [11]. 2D and Kinect features are utilized to represent the various orientation of the human posture and height of the spine using image coordination along with energy, silence ratio are utilized from the frequency domain. Space-time interest points are used to detect specific actions. In addition, 3D skeleton points are employed to recognize falls from the regular activities. Chua et al. [12] proposed a method to detect falls based on human silhouette variation where only three points are used to signify a person posture as a replacement of the conventional bounding box. For fall detection, two lines are formed from the three points and features extracted from these lines are used to identify shape change in human silhouette. Doukas et al. [13] proposed a system for fall detection, which combines both accelerometer motion and visual information capture from 2D camera. The motion information obtained through on-body wireless sensors and body posture features obtained by visual tracking. Finally, SVM is utilized to train a model for fall detection.

Motion information is modeled [14] using hidden Markov models (HMM) along with audio to identify the falling event. The impact sound of the falling person is used as the additional information to avoid false alarms. Standard camera is used in [15] for fall detection. But, the object trajectories are difficult to identify. Hence, vision based fall detection system is proposed in [16] for elderly and patients at home and in health-care centers. It uses an omni-directional camera to evade blind spot. The angle and length variation associated with the body line and motion history images are utilized. Falls are detected using a simple thresholding and decision trees are adopted. Projection histogram of the segmented silhouette, temporal changes in head position is utilized to solve the problem of fall detection. The features are fed to MLP neural network for specific classification and determination of fall event. This approach identified the type of the fall event, like forward, backward or sideward. Authors in [17] used a related technique by employing the k-Nearest Neighbor (kNN) algorithm and evidence accumulation technique to infer human postures for fall detection. Furthermore, they used the speed of fall to discriminate real fall occurrence and an event where the person is simply lying down.

Robustness of fall detection is influenced by occlusion in the presence of multiple objects. Occlusion is the main reason for performance degradation in fall detection systems. Under an occlusion scenario, the objects happen to overlapped and may be found moving together in a scene. Occlusion can be classified as complete, partial and non-occlusion. Handling occlusion is a crucial concern in fall detection and is required all through the occlusion. During occlusion, initially, while two foreground objects occlude each other, the foreground blob of both the objects will form a single blob and turn out to be a difficult issue to discriminate the pixels of both the objects precisely. Secondly, at the time of occlusion, the real position of a traced object becomes hard to make a decision since the visibility of the object becomes entirely missing. Further, it is intricate to recognize the reappearing object after occlusion. The issue becomes more mystifying predominantly while tracking several objects having identical appearance. Complete occlusion during tracking is normally handled by fusing the object motion with the appearance model by linear dynamic or nonlinear dynamic models for predicting the location of the reappearing object after occlusion.

The focus of this chapter is to explore the possibilities to detect a fall event by observing a short video in real-life situations such as home environment, coffee shop, office and lecture hall by accepting the occlusions of furniture and shadows. The chapter contributes by providing a novel and robust technique for fall detection on single persons under partial occlusion. Motion projection profiles are formulated and the novel methodology is explained to solve the problem by effectively estimating the activity in real-time. This is followed by directions for future research and conclusions. Finally, references are listed at the end of the chapter.

2 Visual Information Representation and Fall Detection Approaches

Conventionally, activity recognition techniques have centered on the branch of pattern recognition and machine learning. The literature classifies the approaches as sensor based, vision based, data-driven and knowledge-driven. In sensor-based techniques, the wearable sensors are attached to an object under surveillance. These techniques employ the sensor network technology for activity sensing and monitoring. The vision-based approaches use video cameras to monitor the behavior. The data will be in the form of digitized video sequences. These approaches make use of computer vision procedures to analyze the visual patterns for recognition. Data-driven approaches are based on probabilistic classification and are competent of managing uncertain information. However, this approach requires huge datasets for training. Activity models constructed using the knowledge based approaches are generally utilized for activity prediction through formal reasoning. These approaches are easy to start, semantically understandable and logically well-designed but cannot handle uncertainty.

For recognizing an activity, still certain improvements are to be done in the existing approaches, particularly to deal with the posture and clothes, moving background, imperfect occlusion, camera motion, zooming, etc. In order to overcome these difficulties, the proposed approach focus on motion information extracted from the video sequences. A video sequence is given as a function $v : R^2 \times R \to R$. A sampling point $(x, y, t, \sigma, \tau)^\tau$ is located in the video sequence at $(x, y, t, \sigma, \tau)^\tau$. The characteristic spatial and temporal scales are σ and τ respectively. The spatial scale (σ) describes the similar structure seen in the image plane. The temporal scale (τ) models similar motion seen over a different length of time. σ and τ establish the spatial and temporal neighborhood size of the descriptor at position (x, y, t).

Traditionally, vision-based fall detection techniques have centered on the branch of pattern recognition and machine learning. These approaches use video cameras to observe the event. The data will be in the form of digitized video sequences. These approaches make use of computer vision procedures to analyze the visual patterns for recognition. Projection profiles that gives a compact representation of motion information extracted from the difference image from the two consecutive frames at time t and $t + 1$ is proposed in this chapter.

The general architecture of the fall event detection system is shown in Fig. 1. Fall detection approaches are normally based on motion identification from video. The following discussions briefly explain the approaches for motion detection and image feature representation approaches. Prior to an in-depth discussion on fall detection, it is helpful to discriminate human activities at various levels of granularity. Activity refers to the intricate sequence of motion patterns performed in a restricted style. For example, the action for the phrase 'take the cup', indicates the sequence of motion patterns such as going near the table, and taking the cup. Similarly, recognizing a fall event originally involves classifying the normal and fall event. Thus the fall event detection approaches naturally follows a hierarchical pattern. Lower level involves

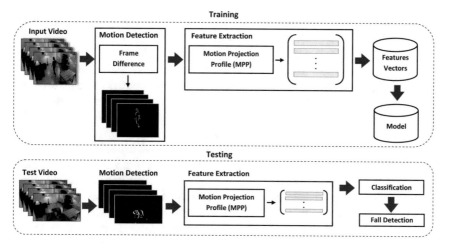

Fig. 1 General architecture of fall detection approach

human detection, tracking and recognition. Higher level semantics of the recognized activities are encoded based on the lower level primitive information.

3 Fall Detection Framework - Overview

For better understanding of the underlying study of this chapter, a real life scenario for fall event detection is experimented. Dataset comprising of activities performed by several actors in different locations such as home environment, coffee house, office and lecture hall is used for experimental purpose. The proposed work employs ideal features extraction approaches and classifiers which reveal promising outcomes.

The overview of the proposed approach is shown in the Fig. 2. The videos are smoothed by the Gaussian filter with a kernel of size 3×3 and variance $\sigma = 0.5$. It is essential to pre-process all video sequences eliminate noise for fine feature extraction and classification. Frame difference is applied to extract the motion information. Motion projection profiles are constructed that exploit the extracted motion infor-

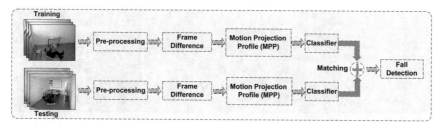

Fig. 2 Overview of the proposed fall detection approach

mation. Support vector machines, k-NN and decision trees are employed to classify the test video to detect the fall event based on the trained model.

The Motion Projection Profile (MPP) features are extracted from the difference image (temporal difference), representing various levels of an individuals posture. The motion projection profile method finds out the measure of moving pixel of each row, column and diagonal (left and right) pixels of the difference frame and they give sufficient motion information to recognize the instantaneous posture of the person as discussed in feature extraction section. Then extracted motion projection profiles of row, column, left diagonal and right diagonal features are resampled to a fixed number of bin (8, 16, 24, 32, 40) values are concatenated into a distinctive vector, in order to obtain a n-dimensional feature vector for each frame in an fall activity video sequence. The extracted features are modeled by SVM (polynomial and RBF kernel), K-NN and Decision tree for fall detection. In this work, fall detection dataset [18] is used in order to evaluate the effectiveness of proposed feature extraction method.

3.1 Feature Description

The extraction of discriminative feature is the most vital problem in fall detection, that represents the significant information that is essential for further study. To identify the person motion across a sequence of images, the current image is subtracted either by the previous frame or successive frame of the image sequences called as temporal difference. The difference images are obtained by applying thresholds to eliminate pixel changes due to camera noise, changes in lighting conditions, etc. This method is extremely adaptive to detect the motion region corresponding to moving objects in dynamic scenes and superior for extracting momentous feature pixels. The temporal difference image obtained by simply subtracting the previous frame t with current frame is a time $t + 1$ on a pixel by pixel basis.

The extracted motion pattern information is considered as the Region of Interest (ROI). Figure 3a and b illustrate the consecutive frames of the fall detection dataset. The resulting difference image is shown in Fig. 3c. $D_t(x, y)$ is the difference image, $I_t(x, y)$ is the pixel intensity of (x, y) in the tth frame, h and w are the height and

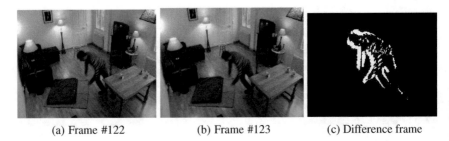

(a) Frame #122   (b) Frame #123   (c) Difference frame

Fig. 3 **a** and **b** Two consecutive frames from fall activity, **c** Difference image of (**a**) and (**b**)

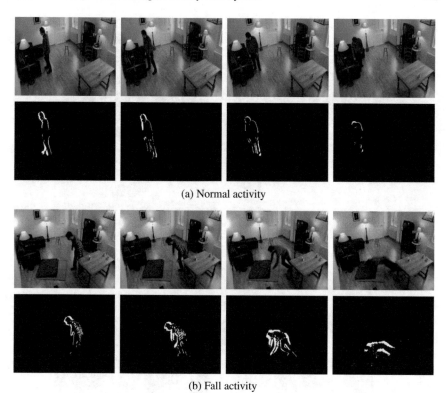

(a) Normal activity

(b) Fall activity

Fig. 4 Sample motion frames from fall detection dataset **a** Normal activity **b** Fall activity

width of the image correspondingly. The sample difference image for normal and fall activities are shown in Fig. 4. Motion information D_t or difference image is considered using

$$D_t(x, y) = |I_t(x, y) - I_{t+1}(x, y)|; \quad 1 \leq x \leq w, 1 \leq y \leq h \tag{1}$$

$$T_k(x, y) = \begin{cases} 1 & if \quad D_k(x, y) > t \\ 0 & Otherwise \end{cases} \tag{2}$$

To recognize the fall activity, motion information is an important signal generally extracted from video sequences. Projection profiles are compact representation of images, since much valuable information is retained in this projection. The projection profiles are extracted from the difference image that consists of motion information only. The row (horizontal), column (vertical), left diagonal and right diagonal projection profiles are obtained by finding the number of white pixels for each bin in a row (horizontal), column (vertical), left diagonal and right diagonal four directions respectively. Figures 5 and 6 shows the measure of variability using mean, variance

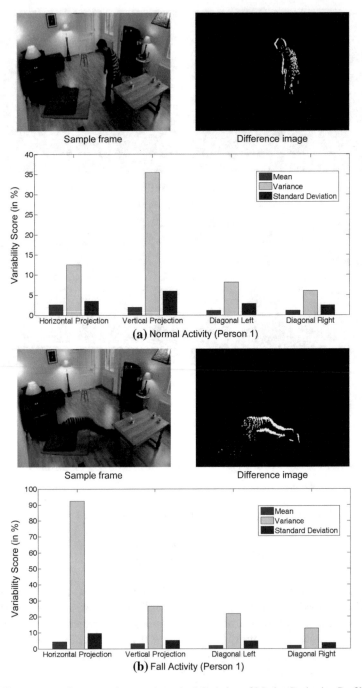

Fig. 5 The measure of mean, variance and standard deviation of Motion Projection Profile of the difference image for normal and fall activity (person 1) in home environment

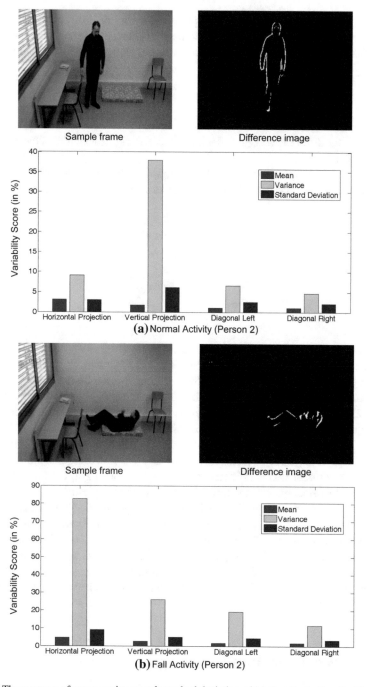

Fig. 6 The measure of mean, variance and standard deviation of Motion Projection Profile of the difference image for normal and fall activity (person 2) in office room environment

and standard deviation to represent the spread of the Motion Projection Profile feature
for normal and fallen activity. The projection $H[i]$ along the rows and the projection
$V[j]$ along the columns of a difference image are mathematically defined by

$$H[i] = \sum_{j=0}^{n-1} T[i, j]; \quad V[j] = \sum_{i=0}^{m-1} T[i, j] \tag{3}$$

where T is the difference image and n and m are the height and width of the difference
image respectively.

For diagonal projections of an object requires only the area moments and the
position thus computed from the horizontal and vertical projections are defined by

$$A_m = \sum_{j=0}^{m-1} V[j]; \quad A_n = \sum_{i=0}^{n-1} H[i] \tag{4}$$

$$\overline{y} = \frac{\sum_{i=0}^{n-1} i H[i]}{A_n}; \quad \overline{x} = \frac{\sum_{j=0}^{m-1} j V[j]}{A_m} \tag{5}$$

where A_m and A_n are an area of the difference image.

The diagonal projection profile is to compute the index for the histogram bucket
for the current row and column. Let the row and column be denoted by i and j,
respectively. Suppose that the dimensions of the image are n rows and m columns, so i
and j range from 0 to $(n - 1)$ and 0 to $(m - 1)$, respectively. The diagonal projection
will require $n + m$ buckets. The affine transformation should map the upper left pixel
into the first position of the diagonal projection, and the lower right pixel into the last
position for left diagonal projection profile feature. For right diagonal profile feature,
the affine transformation should map the upper right pixel into the first position of
the diagonal projection, and the lower left pixel in the last position.

The motion projection profiles $H[i]$, $V[j]$, $D_L[\overline{x}, \overline{y}]$ and $D_R[\overline{y}, \overline{x}]$ are represented
as four feature vectors $\{f_H, f_V, f_{D_L}, f_{D_R}\}$ to depict the difference frame. Then the
feature set is re-sampled with different bin (8, 16, 24, 32, 40) values. The row, column,
left diagonal and right diagonal projection profile bin values are concatenated into
a distinctive vector, in order to obtain a n-dimensional feature vector. The extracted
features are modeled by the SVM with (polynomial and RBF kernel), K-NN, Decision
tree (J48) classifiers for fall detection.

3.2 Support Vector Machines

Support Vector Machine (SVM) is a popular technique for classification in visual
pattern recognition [19, 20]. The SVM is most widely used in kernel learning

Fig. 7 Illustration of hyperplane in linear SVM

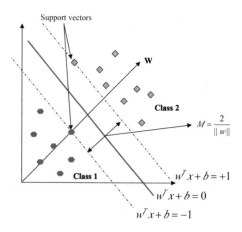

algorithm. It achieves reasonably vital pattern recognition performance in optimization theory [21]. A classification task are typically involved with training and testing data. The training data are separated by (x_1, y_1), (x_2, y_2), ... (x_m, y_m) into two classes, where $x_i \in R_N$ contains n-dimensional feature vector and $y_i \in \{+1, -1\}$ are the class labels. The aim of SVM is to generate a model which predicts the target value from testing set. In binary classification the hyper plane $w.x + b = 0$, where $w \in R^n, b \in R$ is used to separate the two classes in some space Z [22]. The maximum margin is given by $M = 2/||w||$ as show in Fig. 7. The minimization problem is solved by using Lagrange multipliers $\alpha_i (i = 1, \dots m)$ where w and b are optimal values obtained from Eq. 6.

$$f(x) = sgn\left(\sum_{i=1}^{m} \alpha_i y_i \; K(x_i, x) + b \right) \qquad (6)$$

The non-negative slack variables ξ_i are used to maximize margin and minimize the training error. The soft margin classifier obtained by optimizing the functions 7 and function 8.

$$\min_{w,b,\xi} \quad \frac{1}{2}w^T w + C \sum_{i=1}^{l} \xi_i \qquad (7)$$

$$y_i w^T \phi(x_i) + b \geq 1 - \xi_i, \quad \xi_i \geq 0 \qquad (8)$$

If the training data is not linearly separable, the input space mapped into high dimensional space with kernel function $K(x_i, x_j) = \phi(x_i).\phi(x_j)$ [19]. The typical kernel functions are shown in Table 1 where, γ, r, and d are kernel parameters.

In multiclass SVM, N-binary classifier were constructed and one class was separated from all the rest. The training sets of the ith class with positive labels and all others with negative labels. The ith SVM solves i^{th} decision function given in Eq. 6.

Table 1 Types of SVM inner product kernels

Types of kernels	Inner product kernel
Linear	$K(x_i, x_j) = x^T_i x_j$
Polynomial	$K(x_i, x_j) = (\gamma x^T_i x_j + \gamma)^d, \gamma > 0$
Radial basis function (RBF)	$K(x_i, x_j) = \exp(-\gamma\|x_i - x_j\|^2), \gamma > 0$
Sigmoid	$K(x_i, x_j) = \tanh(\gamma x^T_i - x_j + r)$

In RBF kernel, a grid search strategy has been carried out to estimate the best parameter for C and γ in parameter space using LIBSVM [22]. The optimal parameters C and γ are not known before hand, where C - is the slack variable or error weight and γ - is the curvature of the decision boundary or penalty parameter, where these two parameters are used to obtain the optimal classifier performance.

3.3 K-Nearest Neighbor

A K-Nearest Neighbor (K-NN) rule is one of the well-known methods used in pattern recognition. It was initially proposed by Cover and Hart [23]. The K-NN algorithm is a non-parametric classification system has been effectively used in a large number of classification problems for machine learning without having the prior knowledge about the distribution of the data [24]. NN may be a "lazy" learning technique as a result of training information is not preprocessed in any manner. The category allotted to a pattern is that the class of the closest pattern proverbial to the system, measured in terms of a distance outlined on the feature space. The each pattern determines a region by using Euclidean distance measure in feature space called as Voronoi region. The K-NN rule classifies x by assigning the label which most frequently represented among the K nearest samples and the decision is contrived by checking the labels on the K-nearest neighbors. If $K = 1$, then x is simply assigned to the class of that single nearest neighbor.

This algorithm process is based on comparing a given new sample with training samples and resulting training samples that are related to it. It searches the space for the K training samples that are nearest to the new sample neighbors. In this algorithm nearest is defined in terms of a distance metric such as Euclidean distance. Euclidean distance between two samples (or two points in n-dimensional space) is denoted by

$$x_1 = (x_{11}, x_{12}, \ldots, x_{1n}), \quad x_2 = (x_{21}, x_{22}, \ldots, x_{2n}) \tag{9}$$

$$dist(x_1, x_2) = \sqrt{\sum_{i=1}^{n}(x_{1i} - x_{2i})^2} \tag{10}$$

where x_1 and x_2 are two samples with n attributes to measure the distance between x_1 and the point x_2, to take the difference between the consequent values of that attribute in sample x_1 and in sample x_2, where the outcome of a new instance sample is classified based on majority vote of the K - nearest neighbor category. The training samples are described by n-dimensional numeric attributes. Each sample exhibit a point in an n-dimensional feature space.

3.4 Decision Tree

A decision tree is a visual representation of a problem. It helps to decompose a complex problem into smaller parts. This is a formal and structured approach to making decisions [25, 26]. The classification problem of a particular pattern begins at the root node. The root node has different value and it has different links based on the outcome the link followed a subsequent node. The process is continued until the leaf node is reached, which has no further problem. Each and every leaf node having a class label, and the test pattern is assigned class label of the leaf node attained. There are two types in decision tree

1. Classification tree analysis is used when the predicted outcome is the class to which the data belongs.
2. Regression tree analysis is used when the predicted outcome can be considered as a real number.

Classification tree is built through a method known as binary recursive partitioning. This is an iterative operation of splitting the samples into partitions, and then splitting it up further on each of the branches. The operation starts with a training set consisting of pre-classified samples. Each and every possible split is tried and measured, and the best split is the one which gives the largest subside in diversity of the classification label within an each partition [27]. This is continuous for all values, and the victor is chosen as the best splitter for that node. The process is continued at the next node in this manner. An alternate approach to stop splitting is pruning. The process of removing leaves and branches is called as preserving it is used to get better performance of the decision tree when it shift from the training data (where the classification is known) to real-world applications.

3.5 Fall Detection Dataset

The Electronics laboratory, IT and Image (Le2i) created this dataset [18] in 2013 and publicly accessible video for fall event detection. The fall detection dataset captured in realistic environment in video surveillance setting using a single RGB camera with the frame rate of 25 frames/s and having the resolution of 320×240 pixels. This dataset consists of two types of events namely normal daily activities and fall

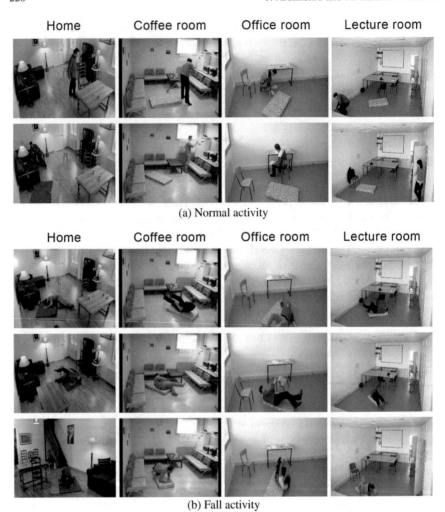

(a) Normal activity

(b) Fall activity

Fig. 8 Example images from fall detection dataset **a** Normal activities **b** Fall events

activity performed by several actors in different locations ("Home", "Coffee room", "Office room" and "Lecture room") various clothing conditions. There is a total of 250 videos sequences whose lengths are around 10 seconds. The video sequence contains typical difficulties like illumination, occlusions and textured background. The actors freely choose position and orientation to perform various normal daily activities (non fall) and falls. The dataset provides the related annotations marking for start and end of each fall event and representing the position of the human body in all frames (Fig. 8).

The experiments are carried out in MATLAB R2013a in windows 7 operating system on computer with Intel Xeon X3430 processor 2.40 GHz with 8 GB of RAM. As

Table 2 Confusion matrix for a fall detection problem

		Actual	
		Fall	Normal
PREDICTED	Fall	True postive	False postive
	Normal	False nagative	True negative

explained in feature extraction section, the n-dimensional MPP features are extracted. The performance of the proposed feature method is tested on SVM (polynomial and RBF kernel), K-NN and Decision tree classifiers using 10-fold cross-validation approach to test the performance of the classifiers.

Table 2 illustrates a confusion matrix for a fall event detection problem having true positive, false postive, true negative and false negative class values. If the classifier predicts correct response of class at each instance and it is counted as "success", if not, it is an "error". The overall performance of the classifier is obtained by error rate, which is a proportion of the errors made over the whole set of instances. From the confusion matrix it is possible to extract a statistical metrics (Precision, Recall, F1-measure and Accuracy) for measuring the performance of classification systems and are defined as follows:

Precision (P) or detection rate is a ratio between correctly labeled instances and total labeled instances. It is a percentage of positive predictions in specific class that are correct. It is defined as in Eq. 11.

$$Precision(P) = \frac{TP}{TP + FP} \tag{11}$$

where, TP and FP are the number of true positive and false positive predictions for the particular class.

Recall (R) or Sensitivity is a ratio between correctly labeled instances and total instances in the class. It have an ability to measure the prediction model and is also called as true positive rate. It is defined as in Eq. 12.

$$Recall(R) = \frac{TP}{TP + FN} \tag{12}$$

where, TP and FN are the numbers of true positive and false negative predictions for the particular class. TP + FN is the total number of test examples of the particular class.

The F-measure is the harmonic mean of precision and recall and it attempts to give a single measure of performance. A good classifier can provide both recall and precision values high. The F-measure defined as in Eq. 13.

$$F_\beta = \frac{(1 + \beta)^2 . TP}{(1 + \beta)^2 . TP + \beta^2 . FN + FP} \tag{13}$$

where β - is the weighting factor, Here $\beta = 1$ that is precision and recall are equally weighted and used to measure the F_β-score which is also called as F1 - measure.

The most common metric accuracy is defined as the ratio between sum of correct classifications and total number of classifications.

$$Accuracy(A) = \frac{TP + TN}{TP + TN + FP + FN} \qquad (14)$$

3.6 Motion Projection Profile with SVM

The extracted motion projection profile (MPP) features are fed to SVM with polynomial and RBF kernels. In polynomial kernel, different degrees (2, 3, 4, 5) were tested. Based on the classification results degree 3 performs better than the other kernel degree. Further, it has been observed that increase in kernel degree doesn't give any improvement in performance. Table 3 shows the classification results of the SVM with polynomial kernel on different bin values. The MPP features are experimented with different bin (8, 16, 24, 32, 40) values and indicates that bin 16 gives good results on proposed MPP feature, when compared to other bin values. Out of the 6910 samples, 2315 (68.64%) samples were correctly classified as true positive (fall activity), 3463 (97.90%) samples were correctly classified as true negative (normal activity), where 1058 (31.36%) samples of fall activity were misclassified as false positive (normal activity) and 74 (2.10%) samples of normal activity were misclassified as false negative (fall activity). In Table 3, Dim refers feature vector dimension.

Table 4 shows the average performance metrics of polynomial kernel, from the results it is clearly indicated that bin 16 gives higher precision = 68.63%, recall = 96.9%, F-measure = 80.35% (trade-off between precision and recall) and accuracy = 83.62% on proposed MPP feature, where high recall value indicate that an SVM with polynomial kernel classifier returned most of the relevant samples correctly.

In RBF kernel, a grid search strategy has been carried out to estimate the best parameter for C and γ in parameter space using LIBSVM. Table 5 shows the classification results of RBF kernel on different bin values. In fall detection dataset the

Table 3 Cross-validated results for SVM using polynomial kernel with different bin (8, 16, 24, 32 and 40) values of MPP features

BINS	Dim*	True postive	False postive	True negative	False negative
8	32	2088	1285	3483	54
16	**64**	**2315**	**1058**	**3463**	**74**
24	96	2236	1137	3431	106
32	128	2236	1137	3418	119
40	160	2227	1146	3398	139

*Dim—Feature Vector Dimension

Table 4 Performance measure results for SVM using polynomial kernel with different bin (8, 16, 24, 32 and 40) values of MPP features (in %)

Bins	Dim	Precision (%)	Recall (%)	F-measure (%)	Accuracy (%)
8	32	61.9	97.48	75.72	80.62
16	**64**	**68.63**	**96.9**	**80.35**	**83.62**
24	96	66.29	95.47	78.25	82.01
32	128	66.29	94.95	78.07	81.82
40	160	66.02	94.13	77.61	81.4

Table 5 Cross-validated results for SVM using RBF kernel with different bin (8, 16, 24, 32 and 40) values of MPP features

Bins	Dim	True postive	False postive	True negative	False negative
8	32	2935	438	3185	352
16	**64**	**2909**	**464**	**3279**	**258**
24	96	2915	458	3228	309
32	128	2717	656	3349	188
40	160	2769	604	3296	241

identified parameter are $C = 5.65$, $\gamma = 0.000481$, where C - is the slack variable or error weight and γ - is the curvature of the decision boundary or penalty parameter, where these two parameters are used to obtain the optimal classifier performance. From the experimental results, bin 16 gives good performance on proposed MPP feature, when compared to other bin values. Out of the 6910 samples, 2909 (86.25%) samples were correctly classified as true positive (fall activity), 3279 (92.70%) samples were correctly classified as true negative (normal activity), where 464 (13.75%) samples of fall activity were misclassified as false positive (normal activity) and 258 (7.30%) samples of normal activity were misclassified as false negative (fall activity).

Table 6 shows the average performance metrics of RBF kernel, from the results it is clearly indicated that bin 16 gives higher precision = 86.24%, recall = 91.85%, F-measure = 88.96% and accuracy = 89.55% on proposed MPP feature, where as precision, F-measure and accuracy values are improved to 5–17%. It is found that the recall value is dropped to 5% when compared to polynomial kernel. Apparently, the overall results give good performance on the RBF kernel.

3.7 Motion Projection Profile with K-NN

In k-Nearest Neighbor classification, the training sample is used to classify each samples of a "target" dataset. An advantage of K-NN classification method is due to

Table 6 Performance measure results for SVM using RBF kernel with different bin (8, 16, 24, 32 and 40) values of MPP features (in %)

Bins	Dim	Precision (%)	Recall (%)	F-measure (%)	Accuracy (%)
8	32	87.01	89.29	88.14	88.57
16	**64**	**86.24**	**91.85**	**88.96**	**89.55**
24	96	86.42	90.42	88.37	88.90
32	128	80.55	93.53	86.56	87.79
40	160	82.09	91.99	86.76	87.77

Table 7 Cross-validated results for K-NN classifier with different bin (8, 16, 24, 32 and 40) values of MPP features

Bins	Dim	True postive	False postive	True negative	False negative
8	32	2825	448	3078	559
16	**64**	**2771**	**602**	**3239**	**298**
24	96	2852	521	3043	494
32	128	2844	529	2882	655
40	160	2847	526	2844	693

its non-parametric form, because it does not build any assumptions on the parametric form of the primary distribution of classes. The extracted MPP features are fed to K-NN (k = 1) for classification. Table 7 shows the classification results of the K-NN classifier on different bin (8, 16, 24, 32, 40) values and indicates that bin 16 gives good recognition results on proposed MPP feature, when compared to other bin values. Out of the 6910 samples, 2771 (82.15%) samples were correctly classified as true positive (fall activity), 3239 (91.57%) samples were correctly classified as true negative (normal activity), where 602 (17.85%) samples of fall activity were misclassified as false positive (normal activity) and 298 (8.43%) samples of normal activity were misclassified as false negative (fall activity).

Table 8 shows the average performance metrics of K-NN classifier, from the results it is clearly indicated that bin 16 gives higher precision = 82.15%, recall = 90.29%, F-measure = 86.03% and accuracy = 86.98% on proposed MPP feature, where as precision, F-measure and accuracy values are improved to 3–13% when compared to SVM with polynomial kernel. But it is relatively less successful to RBF kernel. It is found that the recall value is dropped to 6% when compared to SVM (polynomial and RBF kernel). From the results, it must be noted that the performance of the K-NN classifier is only marginally lesser than other machine learning algorithm.

Table 8 Performance measure results for K-NN classifier with different bin (8, 16, 24, 32 and 40) values of MPP features (in %)

Bins	Dim	Precision (%)	Recall (%)	F-measure (%)	Accuracy (%)
8	32	86.31	83.48	84.87	85.43
16	**64**	**82.15**	**90.29**	**86.03**	**86.98**
24	96	84.55	85.24	84.89	85.31
32	128	84.32	81.28	82.77	82.87
40	160	84.41	80.42	82.37	82.36

Table 9 Cross-validated results for Decision tree (J48) classifier with different bin (8, 16, 24, 32 and 40) values of MPP features

Bins	Dim*	True postive	False postive	True negative	False negative
8	32	2658	715	2772	765
16	64	**2681**	**692**	**2829**	**708**
24	96	2664	709	2814	723
32	128	2674	699	2791	746
40	160	2614	759	2806	731

3.8 Motion Projection Profile with Decision Tree (J48)

The extracted MPP features are fed to Decision tree (J48) for classification. Table 9 shows the classification results of the Decision tree classifier on different bin (8, 16, 24, 32, 40) values and indicates that bin 16 gives good recognition results on proposed MPP feature, when compared to other bin values. Out of the 6910 samples, 2681 (79.49%) samples were correctly classified as true positive (fall activity), 2829 (79.98%) samples were correctly classified as true negative (normal activity), where 692 (20.51%) samples of fall activity were misclassified as false positive (normal activity) and 708 (20.02%) samples of normal activity were misclassified as false negative (fall activity).

Table 10 shows the average performance metrics of Decision tree classifier, from the results it is clearly indicated that bin 16 gives overall precision = 79.48%, recall = 79.11%, F-measure = 79.3% and accuracy = 79.74% on proposed MPP feature. It is found that the overall performance is dropped to 10% on Decision tree classifier, when compared to SVM (polynomial and RBF kernel) and K-NN. From the experimental results, it was found that SVM with RBF kernel is much more efficient to recognize the fall activity with an overall recognition accuracy of 89.55% on the fall detection dataset, which outperforms other machine learning methods.

There is a craving to all of us to "age in place", that is to remain in home and be energetic and self-determining. To achieve this is to use technological support and to use connected smart devices, "Internet of Things" that are becoming a reality. IoT

Table 10 Performance measure results for Decision tree (J48) classifier with different bin (8, 16, 24, 32 and 40) values of MPP features (in %)

Bins	Dim	Precision (%)	Recall (%)	F-measure (%)	Accuracy (%)
8	32	78.8	77.65	78.22	78.58
16	**64**	**79.48**	**79.11**	**79.3**	**79.74**
24	96	78.98	78.65	78.82	79.28
32	128	79.28	78.19	78.73	79.09
40	160	77.5	78.15	77.82	78.44

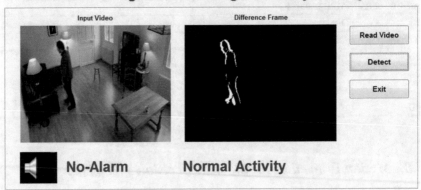

Fig. 9 Snapshot of normal activity detection using motion projection profile

can communicate with one another and with the software running in the cloud. The devices can monitor and can act as sensors to sense the environment and in particular to monitor and assist the elderly people. They also process information and control air-conditioning, locking doors, reminding people to take medications. This chapter proposes an approach for elderly assistance which raises an alarm when a fall event is detected. Figures 9 and 10 shows the detection of normal and fall activity while monitoring elderly people. An alarm is raised as seen in Fig. 10 when a fall event is detected.

4 Future Research Directions

The major application of automatic fall event detection is to identify and provide immediate assistance to the elderly people to avoid some remorseful or dangerous situation. Several potential research directions may still improve the efficiency of the fall recognition system substantially. Video information is a multimodal data,

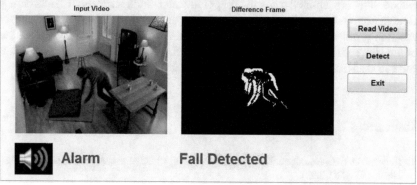

Fig. 10 Snapshot of fall activity detection using motion projection profile

researchers focus mainly on either audio or visual information for their analysis except a few. A combined audio-visual representation may produce an immense rise in the recognition accuracy. Major issue in fall detection is to distinguish such an event from normal sitting or lying down events. Hence, events can normally be split into a sequence of concepts that make up that event. Each one of the concept like objects, actions and audio involved in the incident have their own distinct semantic granularity. Certain issues that are normally seen are (i) how to define/combine these concepts (supervised or unsupervised)? (ii) how to model them? (iii) how to reliably detect them? (iv) how to handle semantic similarity. If these issues could be answered, recognizing fall events would eventually be graceful to determine. Occlusion creates another challenge. Even though optical flow and spatio-temporal techniques are applied to handle this issue, it is always complicated to handle this problem when the object is occluded by another entity with similar shape and motion. Further, privacy of the persons has to be compromised. The area of fall event detection is focused on building a model of the activities performed by the aged people and patients is still open for further research. Another challenge is despite detecting the falls and understanding the basis/causes of falls, predicting a fall event will certainly save and protect the aged people and patients from an injury or further dangerous medical situations.

5 Conclusion

This chapter introduced a new exemplar for human fall event detection. Initially, a novel feature extraction procedure was discussed for fall detection framework. Given the set of video sequences, the motivation is to detect the fall event to enable immediate medical assistance to the elderly and sick by activating an alarm. The problem is formulated using motion projection information and a novel fall event detection methodology is presented using the proposed features. Experimental results demonstrate that the proposed approach outperforms the existing techniques of fall event detection problem. Few of the promising future research directions are discussed towards the end. The authors believe that this effort could afford helpful insights and valuable assistance to the researchers for exploring this topic.

References

1. Rubenstein, L.Z., Josephson, K.R.: The epidemiology of falls and syncope. Clin. Geriatr. Med. **18**(2), 141–158 (2002)
2. Tinetti, M.E., Liu, W.L., Claus, E.B.: Predictors and prognosis of inability to get up after falls among elderly persons. JAMA **269**(1), 65–70 (1993)
3. Abbate, S., Avvenuti, M., Corsini, P., Light, J., Vecchio, A.: Monitoring of human movements for fall detection and activities recognition in elderly care using wireless sensor network: a survey. In: Merrett, G.V., Tan, Y.K. (eds.) Wireless Sensor Networks: Application-Centric Design, INTECH, pp. 1–22 (2010)
4. Bianchi, F., Redmond, S.J., Narayanan, M.R., Cerutti, S., Lovell, N.H.: Barometric pressure and triaxial accelerometry-based falls event detection. IEEE Trans. Neural Syst. Rehabil. Eng. **18**(6), 619–627 (2010)
5. Boissy, P., Choquette, S., Hamel, M., Noury, N.: User-based motion sensing and fuzzy logic for automated fall detection in older adults. Telemed. e-Health **13**(6), 683–694 (2007)
6. Karantonis, D.M., Narayanan, M.R., Mathie, M., Lovell, N.H., Celler, B.G.: Implementation of a real-time human movement classifier using a triaxial accelerometer for ambulatory monitoring. IEEE Trans. Inf. Technol. Biomed. **10**(1), 156–167 (2006)
7. Lin, C.S., Hsu, H.C., Lay, Y.L., Chiu, C.C., Chao, C.S.: Wearable device for real-time monitoring of human falls. Measurement **40**(9), 831–840 (2007)
8. Zhang, T., Wang, J., Liu, P., Hou, J.: Fall detection by embedding an accelerometer in cellphone and using KFD algorithm. Int. J. Comput. Sci. Netw. Secur. **6**(10), 277–284 (2006)
9. Mendulkar, A., Kale, R., Agrawal, A.: A survey on efficient human fall detection system. Int. J. Technol. Enhanc. **3**(4), 96–98 (2014)
10. Willems, J., Debard, G., Bonroy, B., Vanrumste, B., Goedeme, T.: How to detect human fall in video? An overview. In: Proceedings of the International Conference on Positioning and Context Awareness, pp. 1–6 (2009)
11. Planinc, R., Kampel, M.: Introducing the use of depth data for fall detection. Pers. Ubiquitous Comput. **17**(6), 1063–1072 (2013)
12. Chua, J.L., Chang, Y.C., Lim, W.K.: A simple vision-based fall detection technique for indoor video surveillance. Signal Image Video Process. **9**(3), 623–633 (2015)
13. Doukas, C., Maglogiannis, I., Katsarakis, N., Pneumatikakis, A.: Enhanced human body fall detection utilizing advanced classification of video and motion perceptual components. In: Proceedings of IFIP International Conference on Artificial Intelligence Applications and Innovations, pp. 185-193. Springer, US (2009)
14. Toreyin, B.U., Dedeoglu, Y., Cetin, A.E.: HMM based falling person detection using both audio and video. In: Proceedings of the International Workshop on Human-Computer Interaction, pp. 211–220. Springer, Berlin (2005)
15. Nait-Charif, H., McKenna, S.J: Activity summarisation and fall detection in a supportive home environment. In: Proceedings of the 17th International Conference on Pattern Recognition, vol. 4, pp. 323–326 (2004)
16. Miaou, S.G., Sung, P.H., Huang, C.Y.: A customized human fall detection system using omni-camera images and personal information. In: Proceedings of 1st Transdisciplinary Conference on Distributed Diagnosis and Home Healthcare, pp. 39–42 (2006)

17. Nasution, A.H., Emmanuel, S.: Intelligent video surveillance for monitoring elderly in home environments. In: Proceedings of IEEE 9th Workshop on Multimedia Signal Processing, pp. 203–206 (2007)
18. Charfi, I., Miteran, J., Dubois, J., Atri, M., Tourki, R.: Definition and performance evaluation of a robust svm based fall detection solution. In: Proceedings of 8th International Conference on Signal Image Technology and Internet Based Systems, pp. 218–224 (2012)
19. Cristianini, N., Shawe-Taylor, J.: An introduction to support vector machines and other kernel-based learning methods. Cambridge University Press, Cambridge (2000)
20. Mitchell, T.M.: Machine Learning WCB. McGraw Hill, Boston (1997)
21. Vapnik, V.N.: An overview of statistical learning theory. IEEE Trans. Neural Netw. 10(5), 988–999 (1999)
22. Chang, C.C., Lin, C.J.: LIBSVM: a library for support vector machines. ACM Trans. Intell. Syst. Technol. 2(3), 27 (2011)
23. Cover, T., Hart, P.: Nearest neighbor pattern classification. IEEE Trans. Inf. Theory 13(1), 21–27 (1967)
24. Fayed, H.A., Atiya, A.F.: A novel template reduction approach for the nearest neighbor method. IEEE Trans. Neural Netw. 20(5), 890–896 (2009)
25. Duda, R.O., Hart, P.E., Stork, D.G.: Pattern Classification. Wiley, New Jersey (2012)
26. Quinlan, J.R.: Simplifying decision trees. Int. J. Man-Mach. Stud. 27(3), 221–234 (1987)
27. Yuan, Y., Shaw, M.J.: Induction of fuzzy decision trees. Fuzzy Sets Syst. 69(2), 125–139 (1995)

Challenges, Issues and Applications of Internet of Things

Madhvaraj M. Shetty and D.H. Manjaiah

Abstract In world millions of things, objects sense, collect data, share, and communicate these data with each other. Further on analysis, the results obtained are used to initiate some action of decision making and planning in business. This is called Internet of Things (IoT). It has become more relevant to the existing practical world of today due to the evolution of chips, sensors, mobile devices, embedded and pervasive communication, data analytics, and cloud computing. By installing tiny short range of mobile transceivers into everyday items enables new forms of communication and adds new dimension to the world of information and communication. This results to come out new applications and address challenges to the society, such as remote health monitoring, tracking of assets and products, cost savings, optimizing resource usage, enabling quick responses to disasters, smart offices and homes, assisted living, enhanced learning, and e-health. All these will play a leading role in the future. There are numerous potential applications of the IoT, spreading practically into all areas. This chapter throughs light on some potential applications of Internet of Things with its advantages and disadvantages.

1 Introduction

The Internet of Things (IoT) is a futuristic technology where millions of things, objects can able to sense, share information, communicate with each other. All these are interconnected over private or public internet protocol (IP). These objects collects data regularly, analyzes and these results are used to initiate some action and it provides good management, planning, and decision making in different stages of business. The concept of Internet of Things was first coined in 1999 by a member of Radio Frequency Identification (RFID) development community. In turn it has

M.M. Shetty (✉) · D.H. Manjaiah
Department of Computer Science, Mangalore University, Mangalore, India
e-mail: madhvarajj@gmail.com

D.H. Manjaiah
e-mail: manju@mangaloreuniversity.ac.in

© Springer International Publishing AG 2017
D.P. Acharjya and M. Kalaiselvi Geetha (eds.), *Internet of Things:
Novel Advances and Envisioned Applications*, Studies in Big Data 25,
DOI 10.1007/978-3-319-53472-5_11

become more relevant to the practical world of today due to the evolution of chips, sensors, mobile devices, embedded and pervasive communication, data analytics, and cloud computing. The convergence of effective wireless protocols, enhanced sensors, cost-effective processors, and a number of established start ups companies developing the essential applications and management software has eventually made the conception of IoT in mainstream [1]. By installing tiny size, short range mobile transceivers into everyday items enables new forms of communication between objects and people, and between objects. For all these, IoT would become a new dimension to the world of information and communication. These programmed objects has RFID communication technology, embedded technology, wireless technologies, and even quick response (QR)codes. All these objects connected within the current framework of the Internet and uniquely identified by its IP address. They respond to changes in humidity, sound, pressure, weight, temperature, time, motion, light etc., and take the necessary action that they are programmed to.

Internet of Things is a concept and a paradigm that considers environment where different types of objects with unique addressing schemes communicates each other with the help of wired or wireless connections and cooperate with other objects. Also, it creates new services and applications and moves toward accomplishing common goals. The Internet of Things lets people and things to be connected at anyplace, anytime, with anything ideally using any wire or wireless network. The IoT is a symbiotic interaction among the physical, and digital worlds where each individual physical entities can sense, interact, exchange information with each other. However, there are many challenging issues still need to be considered and addressed, in both social as well as technological domains before the concept of IoT becomes a reality completely.

The main issues are to be considered are higher level of smartness by enabling their autonomous transformation and behavior, full interoperability between the objects which are inter connected, security, trust and privacy of users data, efficient utilization of resources. In future, communication, transport, computation, storage and other services will be highly persistent and distributed among people, machines, smart objects, platforms and the surrounding space. It will create a highly decentralized united pool of resources which are interconnected by a dynamic network of networks. In the last few years the growth of markets, applications and their economic potential, impact in addressing societal challenges for the next years has been changed significantly. These societal trends are not limited to transport and mobility, health and wellness, security and safety, communication and e-society, energy and environment. These trends create significant opportunities in the markets of medical applications, consumer electronics, communication systems, automotive electronics etc. All these factors lead to smart life [2].

The reminder of this chapter is organized as follows: Sect. 2 discusses general architecture and various important applications of IoT, Sect. 3 lists various advantages and limitations of IoT. Section 4 introduces implementation challenges and issues related to IoT. Finally, future research direction and conclusion is discussed in Sects. 5 and 6 respectively.

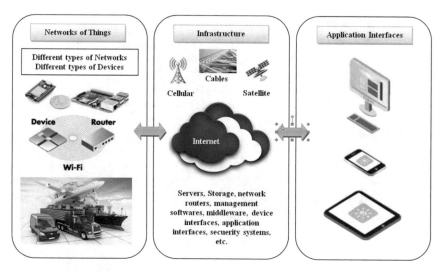

Fig. 1 General architecture of any IoT Application

2 General Architecture of IoT

Architecture of internet of things can be defined in three layers as depicted in Fig. 1. Network of things, internet infrastructure, and the application interface for analysis, are the three key building blocks of any IoT application [3]. Together these are envisioned as foundations for emerging IoT applications.

2.1 Applications

The potentials offered by the IoT make it possible to develop number of applications based on it, of which only a small number of applications are currently deployed. Figure 2 depicts an overview of these applications. With respect to these applications, smart homes and offices, assisted living, enhanced learning, e-health, smart transportation systems, smart factories are few examples of potential and promising application scenarios in which the new paradigm IoT will play a leading role in the future [2, 4–6].

2.1.1 Smart Cities

The IoT can be used to monitor the vibrations of buildings, bridges, and monuments in case the building material is threatened or overloaded. Noise pollution can be controlled around hospitals and schools. It can be used to manage traffic especially

Fig. 2 Applications of internet of things

during traffic jams, peak hours, accidents, and rains. It can be used to manage street lights automatically switch off in the presence of sunlight and switch on at the onset of darkness. Another good application is alerting the officials to empty the trash bins when filled with waste.

2.1.2 Medical and Healthcare Industry

IoT will have large number applications in medical and healthcare sector, with the help of using the cell phones having RFID sensor capabilities as a framework for examining various medical parameters and for delivery of drugs. The gain of IoT increased because of easy monitoring and prevention of diseases, instant medical attention in cases of emergencies, accidents and adhoc diagnosis. Implantable, self configurable, addressable wireless devices is used to capture and store health records that can save a patients life in some emergency situations, particularly for people with heart disease, diabetes, cancer, cognitive impairments, coronary stroke, seizure disorders etc. The patterns of heart pulse rate, blood pressure, digestive system can be continuously monitored and diagnosed for abnormalities. This information can also sent to the doctor in times of emergencies for further analysis. This system will be very useful to senior citizens and disabled people who live independently.

Monitoring Elderly Family Members: This application of IoT creates freedom for the elder peoples to move around safely in outdoors and family members being able to monitor, track their whereabouts. In some situation, elder people may lose their way and were unable to recognize surroundings to remember their way back to home. But using this application, tiny piece of wearable equipment attached to the elderly can helps them to reach home. This equipment will be embedded with location based sensors to report the geographic location and paths that the person has traveled. It can also able to send signals to inform other family members if the person moves away from programmed paths. Simultaneously, it detects variation in their daily routines. User interface can be integrated to this to track the location online by using mobile interfaces.

Continuous Patient Monitoring: This application similar to Elderly Family Member Monitoring. But, it requires the medical services to take care of some operations. It requires the use of medical sensors installed on body of the patient to monitor vital body conditions of the patient such as temperature, sugar levels, and heartbeat. The application examine the current condition of the patients health for any variation, abnormalities and it can predict if the patient is going to face any health problems. It performs predictive analytics, complex event processing (CEP) used to derive information to compare against the existing statistics to make a judgment.

Smart Pills: These are ingestible sensors that are swallowed and it records various physiological measures of a human body. It can also be used to confirm, that a patient has taken his prescribed tablets, and can measure the effects of the medication in the body.

2.1.3 Retail, Logistics and Supply Chain Management

Internet of Things provides several benefits in retail, logistics and supply chain management(SCM) operations. For example, RFID-equipped products and smart shelves can able to track present items in real time and retailer can optimize many applications according using these things, such as, automation of checking of goods receipt, real time monitoring of stocks in warehouse, tracking out-of-stock items, detection of theft/missing products. IoT can provide a huge savings potential in retail by avoiding loss happens when item shelves go empty and customers returns without getting their desired items. Further, these data from the retail stores can made available for optimizing the management of the whole supply chain. That is, manufacturers can produce the right quantities of products to avoid the situations like over-production and underproduction, if they know the sales and stock data from retailers.

The logistic processes from supply chains can make benefit from exchanging RFID data, it can be improved based on the availability of dynamic and fine-grained data gathered in the real world directly by some of the IoT applications, such as individual items, trucks, shelves, pallets etc. In shops, IoT can offer various applications such as fast payment solutions (automatically check-out using bio metrics), guidance according to pre selected shopping list, personalized marketing and detection of potential antigen in a given product, etc.

Shopping assistants: In retail sector, shopping assistants used to provide products recommendations based on customer preferences and to locate suitable products. At present, this is not implemented yet in most of the shopping malls, they do not make any product recommendations for the buyers. The application by IoT can reside in the shopper's mobile devices (phones, tablets) and provides shopping recommendations based on his profile, preferences. Using services that have context aware computing, the IoT application captures data feeds such as locations, ongoing promotions and types of stores (open API or malls websites), next this application attempts to match the buyer's shopping requirements or prompts the user to set any preferences. If the buyer wants to search and locate for a particular product, the application guides that user to destination where the product present from his current location, by using technology such as Wi-Fi, short range sensors.

2.1.4   Manufacturing Industry

By using IoT, manufacturing processes can be automated, optimized, remotely controlled. Inventory and the supply chain can easily managed, can diagnose if the machines require repair and maintenance, can monitor the emission of toxic gases from machines to avoid harm to workers' health and the environment. By linking each items/devices with information technology using embedded smart objects, unique identifiers, data carriers that able to communicate with information system, production processes can be improved and the entire life cycle of items (from production to disposal) can be monitored continuously. By tagging products and its containers, greater transparency can be obtained about its location, disposition, status of the shop floor and status of production machines which are running. Fine grained data from IoT serves as input for logistics and other advanced production schedules. In future, Intelligent manufacturing solutions with self-management can be designed.

2.1.5   Aerospace and Aviation Industry

Internet of Things can help to improve safety and security of different services and products by reliably identifying fake products and elements. In aviation industries, it is vulnerable to the problem of suspected unapproved parts (SUP), which is an aircraft part that does not qualify to meet the provisions of an approved part and does not meets quality constraints of the industry. Thus, SUPs are seriously violates the aircraft security standards. It is possible to solve this problem with help of concept of IoT by introducing electronic chips for certain parts of aircraft, which collects and stores data about their condition, safety, origin and other critical events during its life cycle. By storing these data in decentralized database system as well as on

RFID chips will helps to identify the variation in parameters. By this way, operational reliability and safety of aircraft can be significantly improved.

2.1.6 Automotive Industry

Applications of IoT in the automotive industry include the use of smart devices to monitor and report various actions from pressure in tires to proximity of other vehicles. Streamline vehicle production, improves logistics, quality control and customer services improved by the use of RFID technology. The objects attached to the vehicle parts contains information such as manufacturer name, where and when the product was manufactured, product code, serial number, type and the precise location of the item at that moment in some applications.

Radio Frequency Identification technology provides real time information in the process of maintenance, manufacturing and it offers innovative ways for managing these processes more effectively. Dedicated Short Range communication (DSRC) system will possibly help in reducing interference with other equipment and thus by achieving higher bit rates for communication. V2V (Vehicle-to-vehicle) and V2I (Vehicle-to-infrastructure) communications will certainly makes enhancement in ITS (Intelligent Transportation Systems) applications such as vehicle safety services and traffic management and it will be fully integrated in the IoT infrastructure in the future.

2.1.7 Telecommunications Industry

IoT will create new applications and services by merging diverse tele communication technologies. An illustrative example is the use of GSM, WLAN, GPS, Near Field Communication (NFC), low power Bluetooth, multi-hop networks and other sensor networks together with Subscriber identity module (SIM) technology. In this types of services and applications the reader (i.e., tag, sensors) is a part of the devices like mobile phone, and different applications share the same SIM card. NFC enables communications, transmission of data among devices in a simple with more secure way by having the objects close to each other. The mobile phone can be used as NFC reader and transmit the data to a central server in some applications. Multiple types devices can join the same type of network and perform peer-to-peer communication for specialized intentions; also it can increase robustness of networks and communications channels by optimizing utilization. These devices can form ad hoc peer-to-peer networks in disaster conditions to keep the flow of critical information going in case of access point of infrastructure fails.

2.1.8 Energy Management

Facilities Energy Management: This application is combination of Information systems, Operational Technology (OT) and advanced metering, that is capable of tracking, reporting and alerting operational staff in real time. These management systems are highly capable of allowing dynamic visibility over buildings and other facility performance. They can also provides dashboard view for energy consumption levels, varying degrees of granulation and allows data feeds from a wide range of building equipments and other subsystems.

Home Energy Management: HEM optimizes production and consumption of residential energy. The HEM system includes applications that analyze energy usage levels, and energy management sensors that are connected to home area network (HAN) that responds to the variable power supply when optimizing energy. A combination of these solutions can contributes towards reducing overall energy consumption and carbon emissions from homes. The IoT can also be used to control the appliances in home remotely and useful in detecting and avoiding thefts.

2.1.9 Transportation Industry

IoT offers applications for screening of bags, passengers, commercial carriers, fare collection, toll systems and management of goods at international cargo system that support the governments security policies and to the transportation industry which can meet the increasing demand of security in the globe. Transportation of people and goods can become more efficient by monitoring traffic jams at the path using cell phone devices of the users and deploying of intelligent transport systems (ITS) in it. Transportation companies would become more productive in packing containers, since the containers can itself scan and weigh themselves. Using IoT technologies for managing passengers luggage in airports will enable automated tracking and sorting with increased security.

Accident avoidance: Traffic vehicles can play a leading role in providing better road safety by sensing and monitoring actions of each other on the roads. This application can be programmed and installed on vehicles' on-board equipment (OBE) and with the help of radio sensors installed; this application can warn the drivers of vehicles which may face accidents that lie ahead on the road. For e.g., this application has capacity to interpret a series of complex events occurred such as poor visibility conditions due to heavy rain, slippery roads and strong wind, etc. Further, it can alert or advise the driver on how to drive safely in such complex situations. Sensors with infrared (IR) technology can help to measure the distance between other vehicles or the conditions of the road and feeding this data to application to alert the drivers to avoid and drive clear of a potential accident site.

Distributed Urban Traffic Control systems: This is useful IoT application in todays' world that enables tracking of vehicles locations in real time and provides an appropriate management traffic control system in response to handle road conditions. These control system extremely helpful in the time of emergencies such as setting up of fast lane corridors for emergency services such as ambulances and fire brigades to pass through during heavy traffic conditions.

2.1.10 Pharmaceutical Industry

In Internet of Things paradigm, Pharmaceutical Industries can have many potential benefits by attaching smart labels to drugs, tracking them through its supply chain and monitoring their status with help of sensors, for e.g., drugs requiring specific storage conditions (particular temperature) can be continuously monitored using sensors and alerted if the conditions were violated during transportation. Also application of IoT can be implemented for drug tracking that allows the detection of duplicate products and keep the supply chain system free of frauds. The smart labels on the medicines can also directly benefit the consumers, for e.g., informing consumers about expiration dates, dosages and ensuring the authenticity of the medication. Further with a smart medicine cabinet that collects data transmitted by the drug labels and then reminds patients to take their medicine at appropriate intervals of time.

2.1.11 Environment Monitoring

Using of wireless identifiable devices and IoT technologies in environmental conservation and other green applications are one of the top most promising market segments in the future. There will be an increased usage of these devices in environmental friendly programs in worldwide, like bio-monitoring, remote sensing, soil monitoring, water monitoring and air quality monitoring. IoT can be used to advance environmental programs, including the collection of recyclable materials for the reuse, the disposal of electronic waste (RFID used to identify electronic subcomponents of personal computers, mobile devices and other consumer electronics products to increase the re usage of these sub parts and to reduce e-waste). A very important IoT application is detecting natural calamities and pollution. It can monitor the emissions from vehicles and factories to minimize air pollution, can track the release of waste and harmful chemicals into the rivers and sea, thereby reducing water pollution, can also maintain the quality of water being supplied for drinking. In some critical situations, it can send warnings of tsunamis and earthquakes by detecting tremors, it can keep the water level of dams and rivers under surveillance to be alert in case of floods.

3 Advantages and Limitations

Internet of things makes life simple in many extent. Simultaneously it leads to many advantages and limitations. Keeping in mind the applications, this section highlights the advantages and limitations of internet of things.

1. *Automation of daily tasks*: The Internet of Things allows automate and control the tasks that are done on a daily basis and avoids human intervention leading to faster and timely output. It also leads to uniformity in the different tasks in a day with maintaining the quality of service. It also offers to take necessary action in case of emergencies.
2. *Efficient and saves time*: The machine-to-machine (M2M) interaction provides better efficiency, accurate results that obtained quickly. These results saves valuable time. Instead of repeating the same tasks every day, it enables users to do other creative jobs.
3. *Saves money*: Optimum utilization of energy and resources can be achieved by adopting this technology and keeping the devices under surveillance by alerting in case of possible breakdowns, bottlenecks, and damages to the system. IoT fundamentally proves to be very useful to people in their daily routines by making the devices communicate to each other in an effective manner thereby conserving energy and saving cost.
4. *Better quality of life*: All the applications of IoT technology culminate in increased comfort, better management and convenience thereby improving the quality of life.
5. *Information*: It is obvious that having more information related to one problem helps making better decisions.
6. *Monitor*: The most obvious advantage of IoT is monitoring. For instance, knowing that printer low on ink could save another trip to the store in the near future, monitoring the expiration of products can and will improve safety.

Now, various limitations and brief discussion on it are listed below.

1. *Loss of privacy and security*: As all the industrial machinery, household appliances, public sector services like water supply and transport, and many other devices are connected to the Internet, a lot of information is available on internet. With all of this IoT data being transmitted, the risk of losing privacy increases. The information transmitted and stored is exposed to attack by hackers. It would be very unfortunate if confidential and private information is accessed by unauthorized intruders. As a result, safety is ultimately in the hands of consumers to verify all automations.
2. *Compatibility*: As devices from different manufacturers will be interconnected, the issue of compatibility in tagging and monitoring comes up, this disadvantage may drop off if all the manufacturers agree to a common standard for each equipment, even after solving compatibility, technical issues will persist.
3. *Complexity*: The IoT is a complex and diverse network. As with all other complex systems, there are more chances of failure. Any bugs and failure in the hardware device or software application will have serious consequences. Even simple power failure can cause a lot of disturbance.
4. *Lesser Employment*: With all daily tasks are getting automated, naturally, there will be fewer requirements of human resources. The unskilled workers may end up losing their current jobs in the effect of automation of daily activities. This lead to major issue, unemployment in the society.

5. *Technology Takes Control of Life*: Our lives will be increasingly controlled by the technology, and will be completely dependent on it. Ourselves has to decide how much of daily activities are willing to automated and controlled by technology.

4 Implementation Challenges and Issues

While the current technologies make the concept of IoT feasible, a large number of challenges and issues lie ahead for making the a large scale real world deployment of IoT applications, the following are key challenges [2, 4].

- **Network Foundation**: Limitations of the current Internet architecture in terms of availability, mobility, manageability, scalability and the types of networks are some of the major barriers to IoT. And address restriction, automatic address setup, security functions such as authentication and encryption, and multi cast functions to deliver data efficiently are some issues.
- **Security, Privacy and Trust**: Securing the architecture of IOT (identification arbitrary attacks and malicious software and protection of IOT)
  Requires privacy enhancement technologies, relevant protection laws, methodologies, standards and tools to control over personal information (data privacy) and individuals physical location movement (location privacy).
  Trust has to be a part of the design of IoT and must be built in. It Requires easy and secure exchange of critical, private and sensitive data (because smart objects will communicate on behalf of users/organizations with services they can trust).
- **Managing heterogeneity**: Techniques required for managing heterogeneous environments, applications and devices constitute a major challenge.
- **Information retrieval**: Managing and mining large amount of information to provide useful services.
- **Architecture design**: Designing an efficient architecture for sensor networking and storage.
- **Protocol development**: Developing sensor data communication protocols, that includes publish or subscribe mechanisms, sensor data query and data stream processing mechanisms. Additionally, designing sensor data mining techniques such as aggregation, correlation, filtering, etc.

5 Future Research Directions

There are several areas in which further research is required in making deployment of the IoT more robust, reliable and efficient [3, 7]. Some of the areas are identified as discussed below.

- *In identification technology domain,* further research is required in development of new technologies that address the identity management, global ID schemes, identity encoding/encryption, authentication of third parties, revocable anonymity, repository management, global directory lookup services for IoT applications.

- *In communication protocol domain,* the issues that need to be addressed: communication spectrum and frequency allocation, design of energy efficient communication protocol, multi frequency protocol, to reduce the need of hardware upgrades for new protocols and design of high performance, scalable algorithms and protocols.
- *In architecture design domain,* some of the issues that need attention: interoperability of heterogeneous systems, design of distributed open architecture with end-to-end characteristics, clear flexibility and layering to physical network, neutral access, decentralized independent architectures based on peering of nodes etc.
- *In network technology domain,* further research is required on network on chip technology, considering chip communication architectures for dynamic configurations, parameterized architecture with a dynamic routing scheme and a variable number of allowed virtual connections.
- In addition, *power-aware network design* (that turns on and off the links in response to traffic on demand), *scalable communication infrastructures* (to support the communication among modules based on varying workloads and/or changing constraints) are some of the important research issues.

6 Conclusion

The confluence of advanced sensors, efficient wireless protocols, cheaper processors, and a number of companies developing necessary applications and tools for simple management has finally produced the concept of the Internet of Things. By installing tiny short-range mobile transceivers into a wide array of everyday items and additional gadgets, enabling new forms of communication between objects and people, and between objects themselves, and Internet of Things would add a new dimension to the world of information and communication. It continues to establish its important position in the context of Information and communication technologies and the development of society in the future. Whereas basic concepts and foundations have been elaborated and reached its maturity, further efforts are necessary to unleash the full potential of IoT. We also get an indication of the important aspects that need to be further studied and developed for making large-scale deployment of IoT a reality.

References

1. Atzori, L., Iera, A., Morabito, G.: The internet of things: a survey. Comput. Netw. **54**(15), 2787–2805 (2010)
2. Bandyopadhyay, D., Sen, J.: Internet of things: applications and challenges in technology and standardization. Wirel. Pers. Commun. **58**(1), 49–69 (2011)

3. Gubbi, J., Buyya, R., Marusic, S., Palaniswami, M.: Internet of Things (IoT): a vision, architectural elements, and future directions. Futur. Gener. Comput. Syst. **29**(7), 1645–1660 (2013)
4. Miorandia, D., Sicarib, S., Pellegrinia, F.D., Chlamtac, I.: Internet of things: vision, applications and research challenges. Ad Hoc Netw. **10**(7), 1497–1516 (2012)
5. Bassi, A., Horn, G.: Internet of things in 2020, pp. 1–29. White Paper, Joint European Commission/EPoSS expert workshop on RFID/Internet-of-Things (2008)
6. Kopetz, H.: Internet of things. Real-time Syst. **67**(4), 307–323 (2011)
7. Weber, R.H., Weber, R.: Internet of Things. Legal Perspective, Springer, **230** (2010)

Application of Technologies in Internet of Things

P. Geethanjali

Abstract In this modern smart era, various things in the physical world are connected to the Internet to share information. It is possible to communicate with every things with sensors connected to it. The data from diverse sensors may be collected at regular intervals and collected data may be translated to track objects, control the objects by decision making algorithms, etc. These features of Internet of Things (IoT) revolutionized the medical applications to provide cost effective quality health care with improved efficiency of manpower. The IoT alleviated the obstacles in medical applications in maintaining the records of patients, regular and remote monitoring of data as well as providing automatic diagnosis information to care givers or to the doctors without human intervention. The knowledge of sensors, its connectivity, communication protocol is essential for the neophytes in the development of IoT solutions in health care. Most literature, demonstrates the application of IoT and not much literature describes the various components that are censorious for understanding IoT. Therefore, this chapter covers the various components such as sensors, data collection and communication, software technology that are essential to implement IoT.

1 Introduction

In the near future, the demand for medical assistance will increase due to growing population and elderly people. Further, caregivers and families did not find adequate time to monitor the patient, schedule dose of medicine and managing their health record, look after life style with medication, diet after the treatment during the recovery process and so on. In the world, countries striving hard to provide affordable health services in a more efficient way at the same time without much consumption of resources, time and manpower to the growing demand. The Internet based smart technologies will help to tackle these issues compared to traditional hospital

P. Geethanjali (✉)
School of Electrical Engineering, VIT University, Vellore 632014, Tamilnadu, India
e-mail: pgeethanjali@vit.ac.in; pganjali78@hotmail.com

© Springer International Publishing AG 2017   245
D.P. Acharjya and M. Kalaiselvi Geetha (eds.), *Internet of Things:
Novel Advances and Envisioned Applications*, Studies in Big Data 25,
DOI 10.1007/978-3-319-53472-5_12

treatment [1]. Internet of Things (IoT) [2, 3] is a new revolution of the technology and Internet. Today, several researchers are working on IoT on the development of technologies for medical application (e.g., health care, rehabilitation, surgery). The benefits of IoT are not limited to medical applications. Applying technologies of IoT to industrial systems allows market expansion and diversify new businesses at reduced cost with less barrier. There is a wide range of applications such as utilities, transport, bus systems, defense, etc. where the potential of IoT concepts has been demonstrated. In medical applications, it is crucial to create a real partnership between the human subject and the Internet in order to make potentially usable system to monitor the patients, diagnosing the condition and providing the necessary support. The bottleneck in this case is the integration of different parts such as sensors, control, actuators to the specific requirements of users and challenges related to the on-line monitoring and interfacing data, computational process as well as safety and privacy of shared information. In the recent past, many scientific and technological efforts have been devoted to the creation of human interfacing with electronic devices for exploring the possibility of monitoring and guiding the human from anywhere at any time. This involves seamless linking of interdisciplinary technologies, namely sensors for collection of data, low-power semiconductor devices to process the data, transfer the data, communication technology like wireless transmission and reception, nanotechnology for miniaturization, etc. into the information network to reach the specific requirements of the users. The medical applications using the IoT is one of the promising areas to enable interaction between people and things to sense, monitor, diagnose and provide support to meet the specific requirements of the people anytime, anyplace using any communication protocol.

Following the introduction, basic working principle of IoT is discussed in Sect. 2. Section 3 presents integration of various technologies such as sensor technology, communication technology, middleware technology, cloud computing technology to realize the IoT. Different technologies pertaining to medical applications are discussed in Sect. 4. Chapter concludes in Sect. 5 with a conclusion and future challenges.

2 Basic Working Principle of IoT

The Internet of computers, has evolved into a network of things/objects/devices of all types and sizes, such as smart phones, cameras, medical instruments, home appliances, toys, tablets, home, vehicle, etc. any physical elements containing processor to interact and to share information at any time. Therefore, IoT can be considered as a new paradigm of information and communication technology (ICT), to collect and share information each other with the network for further analysis or processing. There is a plenty of definition for IoT and it is generally defined as "dynamic global network infrastructure with self-configuring capabilities based on standard and interoperable communication protocols where physical and virtual things have identities, physical attributes and virtual personalities use intelligent interfaces and are seamlessly integrated into the information network" [2].

Objects with processor, can interact with the people and other things/objects using sensors. Sensors in the objects, collect data/information and transfer it to the embedded processor. The information from the sensor can be utilized and processed to interpret the hidden information. The information can be communicated to any people/things, anywhere at any time using communication technology. Objects can be localized and communicate information using a global positioning system (GPS) or cellular network, wireless local area network (WLAN). With the advancement of technology, objects can sense and communicate using wireless sensor network (WSN), radio frequency identification (RFID). In addition, objects can communicate via WiFi, Bluetooth, etc. to the network. The identity of the object can be tracked using names and addresses via a universal unique identifier (UUID) [4, 5]. The information communicated from uniquely addressable objects to the Internet can, connect to another object through Internet to interact.

IoT offers greater potential in the medical field application [4, 6], where its primary role is to improve access and quality of care for people living in remote locations at low cost, where people deprived of medical assistance. The regular monitoring of physiological parameters of hospitalized patients require a special attention by those who engaged in monitoring. This can be possible using IoT driven noninvasive monitoring. The sensors collect tangible physiological parameter and communicate and provide access to the expert/caregivers for further analysis in real-time. In this method, gateways and the cloud are used to collect, analyse and store the information wirelessly. It replaces human intervention to monitor the patients at regular basis and facilitate a continuous automated flow of information. In this way, it improves the quality of care and simultaneously can be achieved at low cost by reducing manpower, time who engage in data collection and analysis. The basic block diagram of IoT architecture containing various components is shown in Fig. 1.

Things in IoT may include sensors, actuators, drives, and so on depending on the user service requirement and connect to the base station to provide the connectivity to the various networks. Smart devices equipped with various hardware and software have the limited capability to collect, analyse and store the information for influencing the everyday life. Things like mobile, tablets are equipped with Wi-Fi and Bluetooth

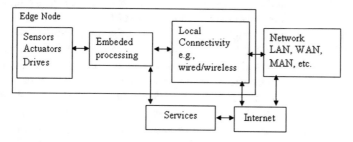

Fig. 1 Basic Block diagram of IoT Architecture

connectivity. These things carry the information and communicate using the mobile network, long term evolution (LTE), 3G/4G to the base station. The base station is linked to wide area network (WAN), which is connected to the Internet. In case of wearable devices such as watch, glass equipped with various biometric sensors sends the data wirelessly to mobile and to base station connected to wide area network, and to the Internet. The data/information collected from the sensor is delivered to the to the cloud via WAN through gateways. Gateways can be designed, depending on service such as clinical application or home automation and may be part of a larger connectivity resource that also manages energy, entertainment and other systems in the home. Subsequently, the data/information may be delivered to the desired destination center via Internet which could be home, office, hospital, etc.

Further, a certain group of people all over the world dont find access to hospital for effective health monitoring and treatment. Medical device designers can also use the IoT platform to access devices for remote monitoring [7]. Also, in IoT, remote monitoring of patients, capturing and storing physiological parameters of patients over time for diagnosis of acute complications from a variety of sensors, analyzing the acquired parameters with necessary intelligent algorithm and securely sharing with medical experts through communication protocols to make appropriate decisions and treatment is possible. In addition, aged-people need attention of the caregiver/ family member during the momentary fall/suffering in a day-today activities when they are alone at home [5, 6]. This is possible via IoT, in which the data aggregation device such as device gateways, home routers, smart phones, etc. captures not only physiological sensor data and other sensor data within the home [1]. The hub captures the various sensor data and stores it in the cloud to access/analyze/process data by those who engaged in caring of the respective individual. Therefore, it is comprehensible that IoT driven solution offers not only monitoring the physiological condition of unhealthy individual, but also to healthy and active people as well. This way of remote monitoring provides timely treatment for proper care of health.

In IoT based healthcare, interconnection of patients, doctors, nurses, services such as hospitals, pharmacies, insurers to provide high quality service to the IoT user is possible through a secure service layer (SSL) that connects to a central command and control server in the cloud [8]. The automation in medical application using IoT, specifically in healthcare eliminates the people from entering the data and the possibility of getting the necessary data in the desired form at any time, in the way the professional expert needs it. This significantly reduces the human error and improves the quality of care for every individual at lower cost [9].

All these are possible through integration with various technologies. The architecture provides the framework for the organization of various components, their relationship among the them and to the environment, operational principles and design. However, a single reference architecture may not be possible for all the applications. For example, sensor based architecture is different from the RFID tag based identification architecture [10]. There are several types of model for interaction, namely things to things, things to human, etc. In addition, the technology of various components is revolutionizing. The next section of this chapter discusses technologies in IoT.

3 Technologies in IoT

Integration of various technologies such as sensor technology, communication technology, information technology, etc. is essential for the effective interaction of things and human for sharing the information and to support in decision making application like diagnosis [11, 12].

3.1 Sensor Technology

In IoT, sensor network plays a vital role and is considered as the system which aggregate and analyze the data from a variety of sensors and share the data to interact. These sensors could be wearable sensors for accurately measuring physiological parameters such as body temperature, heart rate, blood pressure, glucose level, brain activity, etc. or external sensors for sensing, monitoring and analyzing the human activity. External sensors could be cameras at home to monitor the patient fall and smart sensors in pill strips/bottles to monitor the scheduled dose of medication. The sensors used to measure the various parameters could be wired or wireless. There are certain cases where wireless is essential like GPS, RFID to track the user [10, 13, 14].

Now a days, the smart phones are embedded with GPS to identify the location where the user is [10]. However, GPS is expensive due to high energy consumption and not work well in indoors. Therefore, accelerometers are preferable in recent smart phone technology due to low power consumption. Low power consumption is essential in extending battery life that help IoT in medical application as usable as possible. The distance of wireless transmission of data plays a major role in energy consumption. Further, the energy consumption in wireless communication can be reduced by eliminating the continuous raw data transfer using integrated signal conditioning and signal processing capabilities in the sensor at low cost. The sensors are called as smart sensor, which combine a sensor and a microcontroller make it popular and revolutionizing various fields in particular medical applications. Smart sensors with radio frequency waves like RFID and wireless networking-WSN are available in the market to get more and more benefit from IoT.

The concept of RFID [10, 15, 16] has been widely demonstrated in logistics and package tracking which is useful for industries and distributors. This RFID tag consists of a microchip with an antenna to receive the reader signal and transmit the ID using radio waves. In case of passive RFID sensor, any on-board power is not required. The ID is transmitted by harvesting the power of RFID reader. The RFID data transmission range is few meters, i.e., the radio range and may occur in several frequency bands ranging from low frequencies to ultra high frequencies. The on-board powered RFIDs using batteries are referred as active RFID and have larger coverage. In addition to these RFIDs, semi-passive RFIDs are available in the market in which, the RFID is powered while receiving reader signal. In RFID

Protocol version (8 bits)	Data managing organization (28 bits)	Type of Product (24 bits)	Unique ID (36 bits)

Fig. 2 Electronic product code

tags, an electronic product code (EPC) is stored, using 64–96 bit string of data. The distribution of fields in 96 bit EPC of RFID tag is shown in Fig. 2.

3.2 Communication Technology

The networking of different things such as sensors, computer, smart phones, etc. [1, 12, 17, 18]. is essential in IoT to augment interaction to create a smart environment globally with their unique identity. The result is a globally accessible network of things, users, and consumers, who are able to create businesses, contribute content, generate and purchase new services. The cloud enables a global infrastructure to generate new services, allowing anyone to create content and applications for global users. The following section discusses classification of networking to connect to the Internet. Networking may be categorized in many ways. Few methods of categorization are given below.

Network Classification based on things: Networks may be categorized depending on things such as sensor networks, computer network, mobile network, etc. The sensor networking comprises of a number of sensor nodes and one or more sink node. Each sensor node consists sensors, processor, power supply module and communication channel to send the information to the sink. The sensor network consists of sensors, actuators, drives. The purpose of the node is to collect the data, process/analyze using processor and share the results to the other nodes called sinks. There are different types of sensor networking. In this chapter, RFID sensor network and wireless sensor network is discussed.

1. *RFID Sensor Network*: The RFID sensor networks consist of RFID with sensing and computational capability combined with the power of wireless communication using RFID reader. The integration of sensing capability into RFID, develops the wireless identification and sensing platform (WISP) forms sensor node in the RFID sensor network [16, 19].
2. *Wireless Sensor Network*: Wireless sensor network (WSN) [16] is useful in monitoring autonomous devices, distributed spatially using sensor in the wireless communication channel. WSN, consists of a group of nodes connected in certain network topology such as star, mesh, etc. to share the information.

Network Classification based on Mode of Communication: The devices or things can be connected with or without wire. The interconnection of devices using cables is called as wired networks. For example, traditional way of networking of computers using cables to connect, Ethernet ports, USB port to transfer data between connected computers/devices [12, 20].

The interconnection of various devices without using a cable is referred as the wireless network [21]. The wireless network is preferred due to the flexibility in data collection, sharing and access via smart devices, computer and the Internet The wireless network is preferred to avoid the expense incurred due to cable, manpower, etc. and improve the mobility in communication. The interconnection of WSN to Internet, make the IoT as a powerful technology. But, information translation using multiple gateways is essential due to heterogeneous devices, to connect to Internet. Further, the systems are application specific and lacks flexibility due to the requirement of the new protocol in the wireless sensor network. This problem is remedied using IPv6 over low power wireless personal area network (6LoWPAN) [22] to connect the sensor nodes directly to IP protocol. Another important research effort is the data transmission from the sensors and saving energy with the compressed data. The information from a variety of sensors are transferred to the users using various communication standards through gateways [23].

Network Classification based on the Area: Networking can be classified based on coverage area as given below.

1. *Personal Area Network (PAN)*: The PAN is for one user to support communication among devices around him. This may be wired using USB or using devices embedded with Bluetooth, 6LoWPAN, ZigBee, etc. technology.

   - *IPv6 over low power wireless personal area network (6LoWPAN)*: It is a new technology supports IPV6 data packets, large addressing space at low power. 6LoWPAN technology is based on Internet engineering task force (IETF) standard using IEEE 802.15.4 wireless PAN technology. This can be directly integrated to the IPv6 Internet. This technology can interact with each other, connected to different networks and allows communication with other IP-based server or device like Wi-Fi, Ethernet on the Internet. A gateway is required to access the Internet using Ethernet or Wi-Fi. This is an IP-layer gateway and directly connects the device to the Internet.
   - *Bluetooth*: This technology is a successful PAN technology in mobile phones for a short range of communication between mobile phones, computer, etc. This enable networking of Bluetooth wireless devices in the coverage area of about 10 meters. Bluetooth devices operate by emission of signals consisting of unique media access identification number that can be read by Bluetooth sensors in the coverage area. This enables the interaction between the devices embedded with Bluetooth devices. Bluetooth technology can be used to monitor the heart rate, other medical equipment and communicate to smart phone which is indoor. However, Bluetooth 4.0 can be extended to 50 m and operate at a rate of 25 megabits per second (Mbps). Bluetooth also available with different data rates 2.1 and 24 Mbps.
   - *ZigBee*: It is a low-cost, low-power battery powered technology for hoping data in multiple direction. ZigBee network can be connected to the Internet via Ethernet or Wi-Fi. It operates with 802.15.4 protocol standard. In this, it

is possible to implement star network topology, peer-to-peer topology, mesh network topology and cluster tree in case of large scale network. This can be added to the network layer and application layer and possible to establish an isolated network environment. It may be connected to the Internet with application gateway and in parallel TCP/IP stack and application runs over Ethernet or Wi-Fi connect the ZigBee to the Internet [24–26].

2. *Local Area Network (LAN)*: LANs are small area up to few meters, which may be wired or wireless or combination of both. Wireless LANs are referred as WLAN and coverage is up to 100 m using Wi-Fi access. Wi-Fi is a popular network technology integrated into smart phones, tablets, computer and TV for the replacement of wired Ethernet for connecting to the Internet. Wi-Fi connect to the Internet via IP gateway.

3. *Neighbourhood Area Network (NAN)*: The size of NAN is more than 25 Km and it is wireless, e.g. smart metering via ZigBee.

4. *Metropolitan Area Network (MAN)*: The purpose is to cover the city less than 100 km. e.g., high speed wireless Internet access, cable television network, etc.

5. *Wide Area Network (WAN)*: WANs span across a very large geographical area. The Internet is the WAN. In order to improve the usability and communication, graphical user interface based, a tangible display provide an easy approach to access a great deal of information to support effective healthcare assistance through an interaction irrespective type of communication technology. The Internet has revolutionized all the fields and bring the every information in our hand through the interaction of various heterogeneous wide and local area networks. The people can connect to the Internet using the service of Internet service providers (ISP). The Internet service providers, who provide direct Internet access to the end user is referred as local Internet service providers. The local ISPs provide access by connecting regional ISP. Various regional ISPs connect to National ISPs. All national ISPs connect to form international Internet service provider. This is the hierarchy of Internet service access.

Network Classification based on the Transmission Technology: The computer networks are also classified based on transmission technology broadly as shared media broadcast links and switched point to point links. The devices can transmit and receive data, utilizing a certain set of protocols and standards. Protocols are set of rules to communicate data between two entities. Standards allow interoperability of data internationally in the heterogeneous networks. There are two typical model in communication are open system interconnection (OSI), a seven order layer and Internet model, a five order layer. The IoT based communication uses the 4 layered, transmission control protocol (TCP)/Internetworking protocol (IP).

The four layers of TCP/IP communication protocol are the link layer, network layer, transport layer and Application layer. The application layer combines session and presentation layer of the OSI communication model to the application layer, which is based on software. The seven layers of OSI Model is depicted in Fig. 3a whereas 4 layer of TCP/IP protocol is depicted in Fig. 3b.

(a)   **(b)**

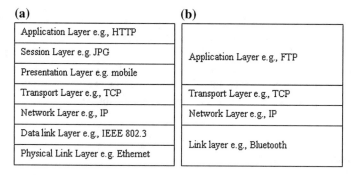

Fig. 3 Models of communication

Physical layer and data link layer are combined to form a link layer to support the network layer in the TCP/IP protocol. This lower link layer is a combination of hardware and software. The network layer packs the data from the lower layer to transport layer as data packets. The transport layer links, the data from the network layer in a form that the user can use in the application layer.

1. *Physical Layer*: The standards of this layer dictate conversion of bits through the physical medium such as coaxial cable, optic fiber, unshielded twisted pair (UTP), wireless for sending and receiving the data. There are various devices in this layer, for instance, RS232, Bluetooth, Wi-Fi and so on.
2. *Data-Link Layer*: The link layers provide standards for data framing for reliable data transfer between two hosts. Some of the protocols for the link layer are Ethernet (802.3), wireless (802.11), LAN, IEEE 802.15, 4e, etc.
3. *Network Layer*: The network layer packs the data with the source and destination address and route the data between the host and through the network. For example, the IP protocol, carry the data packets to the destination device. In this protocol, IP address is provided to the devices to carry the data packet. The other protocols in this layer are Internet control message protocol (ICMP), address resolution protocol (ARP), reverse address resolution protocol (RARP), etc.
4. *Transport Layer*: This layer permits the communication between two device/ network. The most important protocol in this layer is the TCP, which is predominantly used. TCP is useful for interaction of human. UDP is suitable for embedded system based real-time application. Sometime transport layer is referred as stack layer.
5. *Application Layer*: This layer is responsible for the work such as transfer of data, connecting to remote systems, sending and receiving messages, downloading, etc. The protocols in this layer are hypertext transfer protocol (HTTP), simple mail transfer protocol (SMTP), Telnet, remote desktop protocol (RDP), file transfer protocol (FTP), etc.

In communication protocol, the gateway enables the communication between the layers to interconnect heterogeneous application. The gateway can bridge communication, e.g. Ethernet and Wi-Fi, functionality, e.g., Wi-Fi and ZigBee, translation capability IP to 6LoWPAN, TCP to UDP, etc. The different things can connect to each other and to the Internet to share the information through a platform called Middleware. In addition to communication among diverse devices and networks, middleware provides an abstraction/adaptation layer of various domains. This middleware technology is discussed in the next section of this chapter.

3.3 Middleware Technology

The middleware is a software platform in the network layer for integrating various heterogeneous things or networks such as LAN to enable development of different application through information sharing. The role of middleware is a to provide a software platform for communication, hiding details about hardware, operating system, communication protocols, etc. for IoT applications. Some software vendors have developed their own communication middleware like Microsofts WCF. In order to meet the IoT demands, the middleware need to possess the characteristics of interoperability, scalability, abstraction provision, unfixed infrastructure, security and privacy and spontaneous interaction to provide communication among heterogeneous things.

Interoperability: Collaborative communication of heterogeneous things to share the knowledge that serve the different application domain. It can be categorized as network, syntactic and semantic. Network interoperability is responsible for interactive communication of heterogeneous things across the communication network, i.e., connectivity of various layers in the protocol such as the link layer, network layer, transport layer. Syntactic interoperability deals with the data formatting and structure for information exchange. Semantic interoperability provides rules to understand the information which is domain specific. For example, ZigBee protocol does not interoperate with devices having Wi-Fi protocol, if there is no protocol aggregator and translator. The devices are connected to the network mostly using the IP protocol, but this does not satisfy the interoperability requirement at application level. The interoperability is achieved using IoT protocols such as CoAP [7], XMPP, RESTful, HTTP, MQTT, etc. Further, if application layer is not defined, e.g., RFID, Wi-Fi, mobile network, etc. interoperability would not exist, if a protocol exchange is not adopted.

In addition to protocol exchange, interoperability may instill the network interoperation. A gateway provides interoperability between the two protocols and act as a bridge. This middleware support interface between physical layer and application by application programming interfacing (API). The architecture of middleware is layered structure consisting of the fundamental layers for the management module to identify the devices and context, device abstraction for interoperability, syntax and semantic knowledge of devices and interface protocols. There are different APIs architecture namely service oriented architecture (SOA), ubiquitous service-discover service (ubiSD-S), home audio/video interoperability (HAVi) etc. Scalability: Capability of managing scaling up of things to large scale environments.

Abstraction provision: Provision of abstraction at various levels of IoT such as heterogeneous input and output things, interfaces, etc. The device abstraction provides interoperability via different communication interfaces for different application. The application abstraction is responsible for local and remote application.

Spontaneous interaction: Capability to operate spontaneously with the discovery of new device and management of data.

Unfixed Infrastructure: Capability to discover mobility devices in the environment and managing the resources along with the fixed resources from the server.

Multiplicity: Selection of suitable services for communication of things and simultaneous communication with each other. Security and Privacy: Maintaining security and privacy of information in everyday life.

The middleware may be classified based on features, namely interoperating capability-HYDRA, UBISOAP, UBIROAD, SOCRADES, etc. device management-HYDRA, UBISOAP, UBIROAD, SOCRADES, ISMB, ASPIRE, etc. security and privacy-SMEPP, SIRENA, SOCRADES, etc. that suits the interface protocols such as RFID, Wi-Fi, ZigBee, etc. For example, HYDRA middleware can be used for all types of interface protocols. But, ISMB protocol does suit only RFID and sensor protocols and not appropriate for ZigBee, Wi-Fi, etc.

Next, challenging part of IoT is computation capability, memory requirements, especially in advanced services. Some services may require to collect, analyze and process the data and generate control signals. In order to support the various services, cloud technology is essential in IoT. In addition, cloud computing is essential for researchers, healthcare applications in order to communicate and share the knowledge to the deprived local people in national and international level in their respective fields. This concept integrates expertise in different fields to unite and help the community by sharing the various resources.

3.4 Cloud Computing Technology

The hardware/software or combination of hardware and software, connecting the things/sensor, network/sensing domain, to communication networks is referred as gateway. The gateway is the edge of the network for to and fro transfer of data, analyze data to transmit to the cloud e.g. router, proxy server. Cloud computing is an essential platform to share memory, database, software by providing support for devices which do not have/limited memory, computing capability, interoperating capability and network with limited/do not have database through models, namely service models such as software as a service (SaaS), platform as a service (PaaS) and infrastructure as a service (IaaS). The purpose of the SaaS is to support the devices with the required software in order to use the providers applications. The user does not require to have the operating systems, network, storage, etc. for the configuration of application in his device. The PaaS supports a network that does not have a database system for an application. The IaaS support devices and network for providing necessary processing capability whenever required. Therefore, cloud computing provides easy access to, use software, processing platforms and computing infrastructure that is lacking in devices and networks from anywhere through IP network platform using IPv4 and IPv6 protocols. The infrastructure for the cloud is composed of various hardware and software to provide on-demand self service, broad network access, resource pooling, rapid elasticity and measured service.

Another application of cloud in the IoT is machine intelligence based on data mining algorithms to build systems and to automate the process and decision. The concept of removing humans out of the loop improves speed and efficiency in industries and subsequently had an impact on various domains such as business, education and government. However, managing the huge volume of data and interaction among things is challenging. The data management architecture in the cloud consist of three layered structure, namely multitenant database layer, data accessing control layer and business layer. The multitenant layer consists of shared and isolated databases. Shared databases find useful when demand for sharing the data increases. The resource layer, support access to the data and the business layer controls the access of data using web services. Therefore, the hierarchy to access the data is as follows.

Initially Application or service sends a request for accessing the data and then business layers check the rights to access and request RESTful web service to access if approved. Finally, the data is fetched from databases using database fetching connecting layer.

Another challenging part in IoT is security and management of dataflow in addition to the effective integration of various technologies such as sensor, network topology, communication protocols, etc. This enables the requirement of authentication and authorization. Therefore, another challenging technology of identification and authentication needs to be developed to enhance interoperating capability in the world wide. Depending upon the service, the security service will vary. Therefore, gateways are provided with functionalities such as protocol adaptation to bridge different networks, security management of peripheral network, authentication of node for security, etc. It is essential that IoT architecture should be designed to cater the various needs.

4 Technologies of IoT in Medical

In this section, the various medical applications of IoT in particular assistive devices for elder people, assistive living, diagnosis, telemonitoring of physiological parameters, smart hospital, smart emergency applications, etc. [3–7, 11, 12, 20, 27] is discussed to reduce the health care cost and improve the efficiency of man-power. A case study conducted in China and in connection with medical devices is discussed here. Also, various research projects and challenges possessed in health care for IoT is discussed.

Assistive devices and living application: The devices based on RFID tags are used to help visual impaired, hearing impaired, elderly people in a lot many ways to resolve their separation from the social life. RFID tags have been used to find the fall of blind persons or elderly people, obstacle, navigation, etc. by tagging the living/non-living things with RFID. Saaid et al. [22], proposed the application of RFID in navigational system to detect the fall of visually impaired people on the sidewalk by placing RFID at the centre of the sidewalk. Lopez-de-Ipina et al. [17], and Lanigan et al. [27] proposed RFID based navigation in the supermarket with the Bluetooth connection. Shizu et al. [28], proposed a technique that visually impaired person, send the destination location as a voice message and directions are played as voice message using monitoring station. Martin et al. [18], discussed ultrasonic sensor and RFID based obstacle detection system in cane to assist the visually impaired people during navigation. Zhang et al. [29], discussed multiple sensor mounted on shoes to detect obstacles within the proximity of 61 cm ahead. Yelamarthi et al. [15], proposed a RFID and GPS based technology for navigation of visually challenged people. Nicholson et al. [24] proposed a wearable system to get verbal instruction using a handheld computer to assist visually challenged people to navigate, to get the information about the product and the location of the product.

Other than vision impairment, IoT has found potential application to assist the persons with mobility impairment by providing ambient assistive living [3, 9]. IoT for monitoring vitals [6, 15] with things connected in the wheelchair and providing the system to support the users [27] with the IoT technology had been proposed. Further, it is also possible to monitor the falls of people using wearable sensors [28] to alert the caregivers or family member. In addition to management of medical assistance to the impaired people, IoT found useful to locate the medical apparatus and instruments using RFID tags [26], providing necessary data in emergency medical services [15].

A number of research has been conducted for IoT based glucose level sensing, monitoring of ECG, blood pressure, temperature, heart rate, etc. [1, 3, 7, 11–13, 19, 20, 27, 30] with an objective of remote monitoring of patient health and sharing data using IoT devices to an expert to carry out necessary recommendation. Recently, the smart phones are equipped with sensors to monitor vitals such as temperature, heart rate, etc. with the embedded sensors and medical apps [3]. Another, interesting field where IoT find potential application is telesurgery with the help of robotics. But, the main objectives are to achieve a user-centric flexible platform for communication between things, people, things and people as well. Based on IoT objectives, homecare case study for non-communicable disease in China has been studied with an objective of monitoring, servicing and management of information.

Case study on Homecare System: This IoT based system consists of sensors for blood pressure, blood sugar and ECG, cluster of software to store, process and analyse the data and portal for doctors to treat the patients. The caregiver and patients have options to view the medical information. The medical devices are integrated with ARM9 processor [30]. Implementation of hardware is discussed below.

Connection of Medical Devices to implement Homecare system: Physiological conditions are monitored and analysed using microcontrollers. Liu et al. [12] used PIC microcontroller in physiology measuring instrument. ECG and blood pressure measuring units are embedded with PIC. Blood sugar measurement provide via serial port. All these measuring units are connected to ARM processor. But the connection of devices used in clinics is different. The application in the embedded processor collects the data and process the data to the user requirements.

The software has been created to manage profile, history of data for 50 patients. It also process the ECG data with necessary filtering and perform analysis. This has been released in January 2012 and certified in January 2013. The case study was presented for the period of six months. In this study, two doctors serving approximately 3500 patients had been considered. This study reported for 50 patients having hypertension and 27 patients in that group suffering from diabetes as well. Based on this IoT platform, the doctor would be able to give a weekly care to every patient in a group of 3000 patients and improves the efficiency of the doctor.

4.1 Research Projects

There are certain projects realized based on IoT in medical applications for the purpose of monitoring, tracking, assistive living, etc. These research projects include maintaining the database for the purpose of diagnosis to support the doctors in decision making e.g. brain tumour. In order to provide timely support, telemonitoring of physiological parameters at regular intervals are being in use. Information technology sector such as Tata consultancy services (TCS), Wipro, IBM and so on, along with embedded system sector such as Texas instruments, Freescale, etc. is showing great interest in the development of affordable health care for integration of diverse medical devices, maintaining the huge data base, algorithm based diagnosis to support doctors, maintaining data privacy, etc. in a platform based approach.

5 Conclusion and Future Challenges

As the technology for collecting, analyzing, and transmitting data in the IoT continues to mature, it is essential to develop more and more embedded technologies for use throughout IoT driven health care systems. This IoT for medical applications requires active research in the development of miniaturized sensor at low-power and light weight, ability to use with mobile devices and flexible GUI for the user without difficulty, communication protocols, etc. Apart from development of application, it is necessary to make the information transfer secured and save safely with necessary encryption. This is another important module in IoT architecture.

One of the vital aspects to be considered in connection of medical devices is the diversity of devices. The devices could be vital sign monitoring, activity monitoring, safety monitoring or medication monitoring and located at home/clinics. Interfacing the diverse devices, collecting the data and translating the data into useful information is the future challenging task. This could be solved using service-oriented architecture based platform using application programming task [14]. The platform should be capable of comprehending the query and provide the necessary information.

Another, challenge is maintaining data privacy to maintain privacy for the users. Further, the created platform should be scalable with ease of access using graphics user interface. The design of hardware with low power energy consumption is another interesting field of research in IoT.

References

1. Agu, E., Pedersen, P., Strong, D., Tulu, B., He, Q., Wang, L., Li, Y.: The smart phone as a medical device: assessing enablers, benefits and challenges. In: Proceedings of Workshop on Design Challenges in Mobile Medical Device Systems, pp. 76–80 (2013)
2. Vermesan, O., Friess, P., Guillemin, P., Gusmeroli, S., Sundmaeker, H., Bassi, A., Jubert, I.S., Mazura, M., Harrison, M., Eisenhauer, M., Doody, P.: Internet of things strategic research roadmap. In: Proceeding of International Conference on Internet of Things, pp. 9–52 (2013)
3. Dohr, A., Opsrian, R.M., Drobics, M., Hayn, D., Schreier, G.: The Internet of things for ambient assisted living. In: Proceedings of 7th International Conferrence on Information Technology, pp. 804–809 (2010)
4. Jara, A.J., Izquierdo, M.A.Z., Skarmeta, A.F.: Interconnection framework for mhealth and remote monitoring based on the internet of things. IEEE J. Sel. Areas Commun. 31(9), 47–65 (2013)
5. Jih, W., Hsu, J.Y., Tsai, T.: Context-aware service integration for elderly care in a smart environment. In: American Association for Artificial Intelligence Workshop on Modeling and Retrieval of Context Retrieval of Context, pp. 44–48 (2006)
6. Kolici, V., Spaho, E., Matsuo, K., Caballe, S., Barolli, L., Xhafa, F.: Implementation of a medical support system considering P2P and IoT technologies. In: Proceedings of 8th International Conference on Complex Intelligence Software Intensive Systems, pp. 101–106 (2014)
7. Khattak, H.A., Ruta, M., Sciascio, E.D.: CoAP-based healthcare sensor networks: a survey. In: Proceedings of 11th International Bhurban Conference on Applicatioin Science and Technology, pp. 499–503 (2014)
8. Tan B., Tian, O.: Short paper: Using BSN for tele-health application in upper limb rehabilitation. In: Proceedings of IEEE World Forum Internet Things, pp. 169–170 (2014)
9. Shany, T., Narayanan, M.R., Lovell, N.H.: Sensors-based wearable systems for monitoring of human movement and falls. IEEE Sens. J. 12(3), 658–670 (2012)
10. Yang, L., Ge, Y., Li, W., Rao, W., Shen, W.: A home mobile healthcare system for wheelchair users. In: Proceedings IEEE International Conference Computing and Supported Co-operative Work Design, pp. 609–614 (2014)
11. Larson, E.C., Lee, T., Liu, S., Rosenfeld, M., Patel, S.N.: Accurate and privacy preserving cough sensing using a low-cost microphone. In: Proceeding of ACM International Conference on Ubiquitous Computing, pp. 375–384 (2011)
12. Liu, Y., Niu, J., Yang, L., Shu, L.: eBplatform: an IoT based system for NCD patiens homecare in China. In: Proceedings of Globecom 2014-Smbosium on Selected Areas in Communications:GC14 SAC e-Health, pp. 2448–2453 (2014)
13. Tarouco L.M.R., Bertholdo, L.M., Granville, L.Z., Arbiza, L.M.R., Carbone, F., Marotta, M., Santanna, J.J.C.D.: Internet of things in healthcare: interoperatibility and security issues. In: Proceedings of IEEE International Conference on Commununications, pp. 6121–6125 (2012)
14. Xu, B., Xu, L. D., Cai, H., Xie, C., Hu, J., Bu, F.: Ubiquitous data accessing method in IoT-based information system for emergency medical services. IEEE Trans. Ind. Inf. 10(2), 1578–1586 (2014)

15. Yelamarthi, K., Haas, D., Nielsen, D., Mothersell, S.: RFID and GPS integrated navigation system for the visually impaired. In: Proceedings of IEEE International Mid-West Symposium on Circuits and Systems, pp. 1149–1152 (2010)
16. Zhang, X.M., Zhang, N.: An open,secure and flexible platform based on internet of things and cloud computing for ambient aiding living and telemedicine. In: Proceedinfs of International Conference on Computational Management, pp. 44–47 (2011)
17. Ipina, L.D., Lorido, T., Lopez, U.: Indoor navigation and product recognition for people assisted shopping. In: Proceedings of III international workshop of ambient assisted living, pp. 33–40 (2011)
18. Martin, W., Dancer, K., Rock, K., Zeleny, C., Yelamarthi, K.: The smartcane: an electrical engineering design project. In: Proceedings of ASEE north central section conference, pp. 1–9 (2009)
19. Venkatramanan, P., Rathina, I.: Healthcare leveraging internet of things to revolutionize healthcare and wellness, pp. 1–10. White Paper, IT Services Business Solutions Consulting, Tata Consultancy Services Limited (2014)
20. Larson, E.C., Goel, M., Boriello, G., Heltshe, S., Rosenfeld, M., Patel, S.N.: SpiroSmart: Using a microphone to measure lung function on a mobile phone. In: Proceedings of ACM International Conference on Ubiquitous Computing, pp. 111–120 (2012)
21. Shahamabadi, M.S., Ali, B.B.M., Varahram, P., Jara, A.J.: A network mobility solution based on 6LoWPAN hospital wireless sensor network (NEMO-HWSN). In: Proceedings of 7th International Conference on Innovative Mobile and Internet Services in Ubiquitous Computing, pp. 433–438 (2013)
22. Saaid, M.F. Ismail, I., Noor, M.Z.H.: Radio frequency identification walking stick (RFIWS): a device for the blind. In: Proceedings of the Fifth International Colloquium on Signal Processing and Its Applications, pp. 250–253 (2009)
23. Rasid, M.F.A., Musa, W.M.W., Kadir, N.A.A., Noor, A.M., Touati, F., Mehmood, W., Khriji, L., Busaidi, A.A., Mnaouer, A.B.: Embedded gateway services for internet of things applications in ubiquitous healthcare. In: Proceedings of 2nd International Conference on Information and Commununication Technology, pp. 145–148 (2014)
24. Nicholson, J., Kulyukin, V., Coster, D.: ShopTalk:independent blind shopping through verbal route directions and barcode scans. Open Rehabil. J. 9(2), 11–23 (2009)
25. Pang, Z., Zheng, L., Tian, J., Walter, S.K., Dubrova, E., Chen, Q.: Design of a terminal solution for integration of in-home health care devices and services towards the internet-of-things. Enterp. Inf. Syst. 9(10), 86–116 (2013)
26. Puustjarvi, J., Puustjarvi, L.: Automating remote monitoring and information therapy: An opportunity to practice telemedicine in developing countries. In: Proceedings of IST-Africa Conference, pp. 1–9 (2011)
27. Lanigan, P.E., Paulos, A.M., Williams, A.W., Rossi, D., Narasimhan P.: Trinetra: assistive technologies for grocery shopping for the blind. In: Proceedings of the IEEE-BAIS symposium on research in assistive technologies, pp.1–18 (2007)
28. Shiizu, Y., Hirahara, Y., Yanashima, K., Magatani K.: The development of a white cane, which navigates the visually impaired. In: Proceedings of the 29th annual international conference of the IEEE engineering in medicine and biology society, pp. 1–8 (2007)
29. Zhang, J., Lip, C.W., Ong, S.K., Nee, A.: A multiple sensor-based shoe-mounted user interface designed for navigation systems for the visually impaired. In: Proceedings of the fifth annual ICST wireless internet conference, pp. 103–110 (2010)
30. Lopez, P., Fernandez, D., Jara, A.J., Skarmeta, A.F.: Survey of internet of things technologies for clinical environments. In: Proceedings of 27th International Conference on Advanced Information Networking and Applications Workshops, pp. 1349–1354 (2013)

Part IV
Industry and Societal Applications
of Internet of Things

An Appraisal on Human-Centered Internet of Things

A. Geetha and M. Kalaiselvi Geetha

Abstract This chapter outlines the current state of the Internet of Things (IoT) from the people's association point of view. IoT is an intelligent environment where objects become smart and autonomously communicate with one another and human beings, through networks supported by interfaces. The IoT technologies have not only been widely studied in simulation and investigational circumstances but also dealt with real world scenarios. The IoT systems are enhanced by surveying diverse interactions between humans and the IoT to mine the implanted intelligence about individual, environment, and society. The IoT is facilitated by the latest developments in RFID, smart sensors, communication technologies, and internet protocols. In the upcoming years, the IoT is expected to bridge various technologies to enable new applications by connecting physical objects together in support of intelligent decision making. The spotlight of this chapter is to present the support of intelligent human-computer interaction for the IoT and to deal with human-centered concerns in the IoT.

1 Introduction

Nowadays, the internet has changed from a network of interconnected computers to a network of interconnected objects. The dream of the IoT is to create the potential for objects of all kinds including people to communicate with one another by means of the internet. Apart from the technological hurdles such as privacy and trust, currently there is a lack of strong human-centered perspective on the IoT. Modern machines profoundly rely on the cognitive skills of their users, who make decisions and act in real-time. The cognitive skills of human beings include perception, knowledge,

A. Geetha (✉) · M. Kalaiselvi Geetha
Department of Computer Science and Engineering, Annamalai University,
Annamalainagar, 608 002, Tamilnadu, Chidambaram, India
e-mail: aucsegeetha@yahoo.com

M. Kalaiselvi Geetha
e-mail: geesiv@gmail.com

working memory, auditory and visual processing, judgment and evaluation, production of language, reasoning, problem solving and decision making.

The IoT systems [1] usually consist of a set of sensors that collect information, which is then transmitted among different devices without human intervention. But integration of the IoT and human interaction will make the IoT work smarter when input from people is considered. A system that collects data automatically can be compromised by issues of data qualities, such as inconsistent, missing, or unrecognized data which then can result in incorrect analysis. For example, human interaction with the IoT systems through mobile apps can augment automated data collection and correct, complement, extend, or even override the data gathered by the system or the actions it would undertake in response to the analysis of poor quality data. Also the functionalities of mobile apps such as camera, GPS, compass, bluetooth etc. make it possible for a person to collect a wide range of information which can in turn be used to interact with the IoT system.

The IoT systems roll out across industries, such as health management, industrial production, logistics, retail etc. The key to maximizing the usefulness of the IoT systems is to improve the capabilities of the IoT environments by making it easier for real people to contribute to them. The IoT should not be just about alarm clocks that start our coffee maker, or about accessing regular things over the internet but be the ultimate platform for human interaction with the physical world that would turn the internet into a mere medium.

Human-Centered Systems (HCS) are designed to complement human skills [2] and are not the systems that merely replace or automate human activities. HCS tap into our creative abilities that are unnoticed by the conventional problem-solving practices. It depends on human ability to be perceptive, to recognize patterns and to construct ideas that are emotionally meaningful as well as functional. Human-Centered Systems are inspired by behaviors rather than demographics, take place in natural contexts rather than controlled settings, and rely on dynamic conversations rather than scripted interviews. They are empathetic in understanding the needs and motivations of people who make up a community, collaborative in benefitting from multiple perspectives, optimistic in behaving irrespective of the constraints and experimental in making new things possible.

The design of Human-Centered Systems [3] is concerned with integrating the user's view into the software development process in order to achieve a practical system. The significant principles of the Human-Centered systems should include the active involvement of end-users who have knowledge of the system context and should be a mutual process with an active participation of diverse experts with excellent technical skills. In this viewpoint, Sect. 2 describes some of the case studies on human-centered IoT and Sect. 3 discusses the conclusion and future directions of the IoT.

2 Case Studies of Human Centered IoT

This section describes a few of the case studies on IoT. The purpose of this description is to understand how the IoT intelligently help humans survive in vital situations. The IoT can be exercised by humans irrespective of their age and class such as patients, elderly people, disabled adults and children, night drivers etc.

2.1 Drug Tracker for Patient Safety

Patient safety is a major issue for human survival. The principal goals to augment patient care are drug compliance, nominal adverse drug reactions (ADR) and the elimination of medical errors. The drug compliance portrays the degree to which a patient correctly follows medical advice. The ADR results from the combination of two or more drugs. The above are caused mainly due to polypharmacy [4] in hospitals where patients undergo treatment for multiple illnesses. The polypharmacy increases with age and exists in poor countries where some diseases like TB cause the ADR. Such problems can be prevented if the prescribed drugs are examined to find the impediments by electronic prescribing of drugs and maintaining a record of prescribed drugs. An interaction between users and drugs is essential to bring out such drug checking tasks. Antonio J. Jara et al. have proposed a drug checker [5] as shown in Fig. 1 based on the IoT to watch the treatment for drug compliance and to detect ADR. The IoT adopts technologies such as barcodes, radio frequency identification (RFID) and near field communications (NFC) for identification of drugs. Once the drugs are identified, drug fitness according to the patient's profile and personal health record (PHR) has to be verified by means of pharmaceutical intelligent information system (PIIS).

The components of the proposed architecture in Fig. 1 are listed as below:

1. The PIIS is a knowledge-based system which contains a rule engine system to detect the possible interactions between prescribed drugs and ontology where the drug concepts and patients information are described. After the drugs are prescribed by the doctors in hospitals the patients can confirm them before consumption.
2. The database contains a detailed drug description about ingredients and side effects.
3. The drugs are kept up-to-date and indexed by barcode numbers letting for easy and scalable identification.
4. A patient profile is set up, which contains the record of allergies and drug history ontology describes the anatomical, therapeutic, and chemical classification (ATC), dose, drug name, drug interactions, side effects, allergy causing ingredients etc.

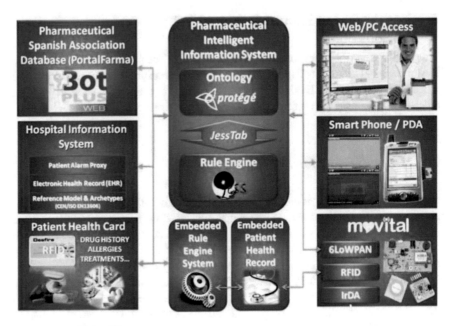

Fig. 1 Drug checker [5]

5. Rule-based system detects drug allergies, skin rashes and breathing problems, drug interactions where one drug changes the effect of another drug, and drug loop where the side effect of one drug is solved by another drug.

The proposed drug checker has adopted many IoT technologies as depicted below:

1. *Personal device based on IoT*: A personal device called Movital based on IoT offering RFID and internet connectivity is developed and offered based on 6LoWPAN. This permits direct patient access to the internet and to PIIS. The RFID lets patient identification and a medium to load the patients' health record by nearing their personal health card to the reader. The drug is identified by integrating RFID tags with drug box that store the drug code as a drug identifier as shown in Fig. 2. The Movital RFID raises an alarm or light signals to remind patients to take their medicines in time. The drawback of this adapted device is its highcost.
2. *Movital IrDA*: Another technology called Movital IrDA as shown in Fig. 3 is attempted in order to reduce the cost of Movital RFID. This tag is designed with low cost IC which consumes less power by transmitting the drug identifier through LED.
3. *RFID/NFC using USB reader*: This technology provides an option to store the patient profiles in the PIIS, and to access their health cards based on RFID.
4. *GS1 barcode using smart phones*: Currently the smart-phones have cameras by which the barcodes of drugs can be scanned and read. The drug ID reaches PIIS with the patient profile using the internet. PIIS checks the drug ID with its

Fig. 2 Drug box with RFID tag [5]

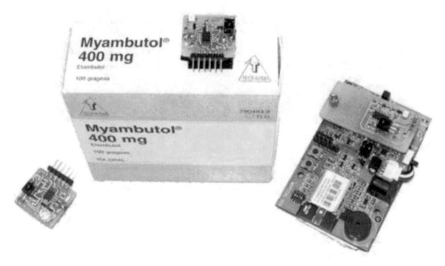

Fig. 3 Drug box with IrDA-based tag [5]

rule-based system and the patient profile and respond with informative details to the phone. This information verifies drug suitability for the patient with a proposed solution. The drug suitability is indicated by the display in green color and unsuitability in red color as shown in Fig. 4.

5. *NFC using smart-phones and pocket PCs*: The Pocket PC or smart phone is placed near the NFC drug tag, which reads the tag and starts the communication with PIIS as shown in Fig. 5.

From the above discussion, it is understood that the proposed system has used three media such as the RFID, the IrDA and the barcode. The proposed solutions

Fig. 4 Barcode reading [5]

have been analyzed and evaluated by a multidisciplinary group of experts and by two pharmacies. Each solution has its own pros and cons. As a result, Movital with RFID is recommended for richer countries whereas Movital with IrDA for low-income countries. The proposed system has made the possibility of utilizing new technologies to improve quality assurance in drug delivery, to improve adherence to drug appropriate consumption, and to reduce clinical errors caused by dosage mistakes and drug interactions.

2.2 IoT Embedded Systems for Disabled Persons

Human-Computer Interaction (HCI) is a challenging task for people with disabilities. Davide Mulfari et al. have examined the use of low-cost embedded systems [6] to support the interaction between disabled users and computers. The main intention is to connect such embedded systems to any computer to help the user work on it without any prior configuration setup so as to adapt to the user disability. These embedded devices are designed in such a way that they manage heterogeneous wireless sensor networks and process data over cloud for IoT purposes. In general, disabled users

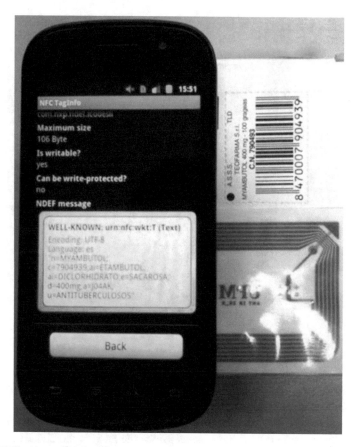

Fig. 5 NFC with smartphone [5]

interact with computers by means of Assistive Technology (AT) equipments. Such AT equipments need special configuration which makes the users difficult to work with their personal AT solutions. This issue can be solved by providing the end user with a smart AT hardware to interface the user with any computer without any installation of drivers or specific applications. As a result, the disabled user is enabled to employ his AT equipment to interact with different computing devices, such as PCs or smart phones. This setup process should make any computer to interact with the HCI system.

The proposed HCI system comprises of Atmega32u4, a microcontroller that supports mouse/keyboard libraries and acts as a resident mouse or keyboard when attached to a generic computer with an USB host port. Another key hardware component of the HCI system is a processor that needs to manage a network connection to access the internet and an USB host port to interface external peripherals. These constraints are accomplished by Arduino Yun board, a microcontroller board based on the Atmega32u4 and the Atheros AR9331 processor which supports a Linux

Fig. 6 Accelerometer-based
sensor node attached on the
head [6]

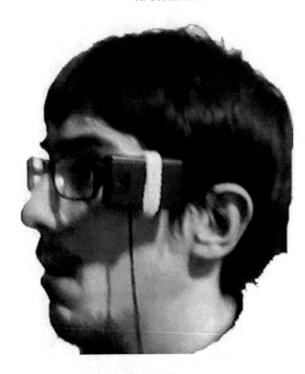

distribution. Yun board is designed with quite a few kinds of sensor node, each sensor node consisting of a dedicated microcontroller, and an interface which enables the communication with a computing device. These sensor nodes are placed on human head or arms to capture their movements. Since the head is the highest body segment, head controls can be considered as alternative for hand controls to control the computer cursor as shown in Fig. 6.

The up and down vertical movements of the head are used for up and down displacement of the cursor while left and right horizontal movements are used for left and right displacement. Hence the disabled users can fully control a computer through head movements given that a sensor node can detect measurement of inclination changes less than $1.0°$. AT is mainly used by persons with disabilities in order to control their computers through resources other than a standard keyboard or pointing device. The other assistive solutions used by some of the related works are as follows:

1. Keyboards with alternative configuration for use with one hand.
2. Electronic pointing devices using ultrasound, infrared beams, eye movements, nerve signals, or brain waves to control the cursor on the screen without use of hands.
3. Sticks worn on the head, held in the mouth or strapped to the chin that are used to press keys on the keyboard.

The aforementioned discussion proves that the IoT devices can be effectively used to achieve human-computer interfaces for users with disabilities.

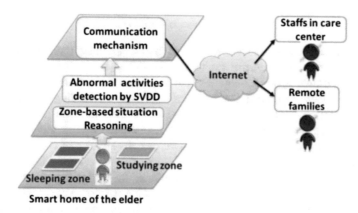

Fig. 7 Architecture of the proposed system [7]

2.3 Elderly People Care Using IoT

Taking care of elderly people plays an important role in the society. Many elderly people are living alone because they are isolated from the society and are deficient in support from their family. Junbo Wang et al. have proposed a new application [7] using IoT to watch elderly people, to grasp the situations around them and to send the information to their care- takers to support them. While watching them, any abnormal activity detection need to be carried out since the services should be immediately offered to them in such cases to avoid any danger.

The proposed technique detects the situations around the user and provides the corresponding services or automatically changes environment to adapt the user. In this proposed work, not only the general situations are detected, but the abnormal activities in the situations are also recognized to support the users. The basic architecture of the proposed means is shown in Fig. 7. Initially the location of the elderly people need to be detected. A special type of sensor network called u- tiles sensor network is built below the floor which captures the location and position relation between the elderly people and surrounding objects based on the zones and other detail information namely status of the home appliances, actions of the user, etc. Then the abnormal activities in the detected situations are recognized by Support Vector Data Description (SVDD) and ultimately, the system sends the corresponding information to care-takers such as remote family and staff in care center.

Following are the abnormal activities need to be detected based on some features as discussed below:

1. Forgetting to take medicine or taking medicines in an incorrect manner can be detected by tracing the duration between taking medicines
2. Going to restroom many times than normal times can be detected by starting time
3. Sleeping for a long time is detected by number of times per day
4. Taking unbalanced and insufficient foods by types of food

5. Inappropriate body exercise by number of times per week

The proposed technique has used a hardware environment called u-tile sensor network to detect the abnormal activities and SVDD for recognizing them. The basic idea of u-tiles sensor network is to detect precise location and position relation between the elderly people and surrounding objects by embedding various sensors, including pressure sensors and RFID antennas under the floor. This network consists of u-tiles under which there are pressure sensors and RFID antennas which are connected with a reader through a PIC based switch. The IDs of users and objects are read by each piece of u-tiles, and the position relation between users and objects are captured. To realize a smart home environment, the u-tiles sensor network need to be set up in the zones where elderly people often have several activities around the bed, in front of sofa, restroom etc. Apart from the u-tiles sensor network, a smart plug has been developed by NTT Company to detect and control the status of home appliances, i.e. turning on or off of a lamp, fan etc.

In the proposed technique four aspects about elders are needed to represent a situation, i.e. zone, status of related home appliances, actions of the elder people and environment. A situation s can be represented by the four tuple, $s = < z, D, A, E >$, where z is used to represent the current zone of the elderly people, D is a set including the status of related devices, A is a set including current actions of the elderly people, and E is a set including environmental factors.

A situation can be detected by framing some detection rules. An example of such a detection rule is "if the person is in the bedroom and if the bedroom light is off and if there is no action" then the situation detected is "sleeping in the bedroom". The next task after detecting situations is to recognize the abnormal activity in these situations. The proposed technique uses SVDD to recognize abnormal activity which constructs a boundary for a dataset to detect target data or outlier. On analyzing the performance of the system by experimenting with 5 subjects the system has totally detected 136 activities out of 146 and achieved an accuracy of 93.2%.

The aforesaid details show the importance of elderly care and prove that the proposed situation-aware system detects abnormal activities of the elderly people on par with the other systems.

2.4 IoT Support for Children Affected by ASD

Autism Spectrum Disorder (ASD) is a serious neurodevelopmental disorder that impairs a child's ability to communicate and interact with others. It also includes restricted repetitive behaviors, interests and activities. Since currently there is no cure for ASD, Ardiana Sula et al. have proposed a novel framework [8] based on Internet of Things (IoT) and P2P technology for supporting children with ASD to help them to interact, respond, and tell parents what they need for their survival.

The IoT is motivated by the victory of RFID technology, which is now widely used for tracking objects, people, and animals. RFID is primarily used to identify objects

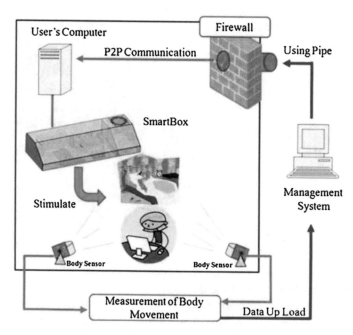

Fig. 8 IoT based framework for supporting children with ASD [8]

from a distance of a few meters and to determine the approximate location of objects provided the position of the reader is known. RFID systems consist of three main components namely RFID tags, the RFID readers, and RFID software. The RFID tag is an electrical device that receives a specific signal and automatically transmits a specific reply. The tag also carries an unique ID data and stores information contents depending on the size of its memory. The RFID reader is a hardware device that is used to read the transmitted data from the tag. The RFID software is a middleware that runs on the RFID reader.

The assistive technologies for the diseased children can be realized through the IoT where the children with autism live in their homes with smart objects, communicating to the outside world in an intelligent manner. The proposed framework is shown in Fig. 8. The system sends information about the children state in real time to therapists using P2P technology and also allows children to interact with other children and parents. The architecture of P2P distributed platform is developed using JXTA technology [9].

The Smartbox device used in the proposed system is integrated as a useful tool for monitoring and controlling children activities. The Smartbox has different sensors such as body sensor, chair or bed vibrator control, light control, smell control and sound control, and remote control socket. They are used to detect body movement, vibrating chair or bed, adjusting room light, controlling room smell, and emitting

Fig. 9 Assistive learning [8]

relax sounds respectively. Such functions relax the children affected by ASD and divert them to concentrate on some tasks.

Assistive technologies, apart from improving the health of the children with autism spectrum disorders, help them in assistive learning. Tagging physical objects to find and analyze data about the object is one way the IoT can be used in education. Figure 9 shows a part of the proposed framework for assistive learning. A child can learn new words through touching the physical objects that are in their vocabulary list. Each physical object would have a RFID tag placed on the item. When this tag is read by a RFID reader or scanned by an application running on a computer or mobile device it would prompt the device to open up a page of information or send a command for an action to happen. The RFID tags can be created and attached by the parents for each of the physical items in the vocabulary list. When the child places the RFID card on the RFID reader, it pronounces the word for the item in their native language. Touching the item will give the child another sense to be engaged and may help them learn new words faster.

According to the above picture about the assistive technologies in the IoT, it is understood that even children affected by ASD can learn and survive on par with the normal children.

2.5 Tracking of Night Drivers for Accident Prevention

Prevention of road accidents during night time is an essential service for night drivers. Most of the road accidents happen due to the drowsiness state of the drivers. Aish-warya S.R et al. have proposed a novel system called Eye Blink Monitoring System (EBM) [10] to alert the drivers when they are under the condition of drowsiness. An embedded system based on psychological state of the drivers is developed to monitor eye and head movements which are useful in warning drivers during their sleepy phase of drowsiness. The sleep state of the drivers is determined by capturing the eye-blink rate using an IR sensor and head movement using an accelerometer. An IOT enabled sensor is used to transmit the entire data collected by sensors over a smart grid network for a quick response to take recovery actions under emergency conditions.

The existing technologies to prevent accidents are not cost effective and not easy to be implemented. Also they are inferior to the proposed technique due to many factors for the reason being that they are not well defined, load elimination is impossible and they respond only after the occurrence of accidents. The main objectives of the proposed framework are as follows:

1. Creation of an eye blink & head movement monitoring sensor system for Drowsiness detection.
2. If drowsiness is detected,

   • Alerting the driver by means of a buzzer
   • Reducing the speed and stabilizing the vehicle.

3. Mediating the sensor information and tracing accident location using GPRS for help and rescue.
4. Displaying the activities of designed system on LCD display.

The block diagram of the proposed framework is shown in Fig. 10. The designed embedded system is interfaced with another mobile phone having an android platform through an IOT application. Such mobile phone offers notification to the host about the status of the embedded system in case of drowsiness and accident occurrence via alarms, text messages and voice notifications.

The proposed work involves measurement of the eye blink using an IR sensor and head movement using an accelerometer. The IR transmitter is used to transmit the infrared rays in our eyes. The IR receiver is used to receive the reflected infrared rays of the eyes. If the eyes are closed then the output of IR receiver is high, otherwise the IR receiver output is low. To know whether the eye is in closing or opening position, the output is provided to a logic circuit for alarm indication and the status will be displayed on LCD. The accelerometer is placed on the driver's fore-head which measures tilting angle of the drivers either towards forward or backward direction and towards left or right direction from the driver's knee. If the tilting angle exceeds a certain threshold range then the output is given to logic circuit to indicate the alarm and the status is displayed on LCD. Figs. 11 and 12 show the sensors and message display on LCD.

Apart from helping the night drivers, the proposed technique prevents the drunkards from drowsing and rash drivers from over speed in driving. The main pros of the proposed work are as below:

1. Effective and easy implementation.
2. User friendly interface.
3. Prevention of over speeding of vehicles.
4. Prevention of accidents before their occurrence.

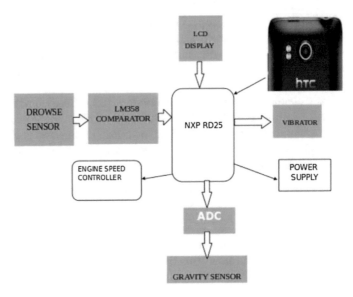

Block diagram of the proposed tracking system [10]

Fig. 11 IR sensor and
accelerometer [10]

5. Load reduction due to the usage of cloud computing in implementation.

The countable demerits are requirement of active internet connection and standard bodies to enable IoT. From the above narration, it is clear that the IoT can revolutionize the way the embedded systems interact and respond for variety of applications especially in case of night drivers by monitoring the state of their drowsiness for a quick, safe and effective response for a safer road travel.

Fig. 12 LCD response on
drowsiness detection [10]

2.6 IoT Based Healthcare System for Wheelchair Users

Healthcare monitoring of residents such as disabled or elder persons has become a focus of recent researches and developments. Lin Yang et al. have proposed a home mobile healthcare system [11] for wheelchair users, based on IoT technologies. The primary ambition of the proposed work is to build an intelligent system with real-time monitor and interaction to take care of wheelchair users at home. The shortcomings of other related technologies have been overcome by the efficient design of the architecture of the proposed work as shown in Fig. 13.

The architecture of the proposed work constitutes three components namely Wireless Body Area Networks (WBANs) with smart objects, smart phone and data center layer. WBANs is a people-centric network for wheelchair users, containing nodes of vital physiological parameters and living environment to detect wheelchair falling and controlling. The purpose of detecting the falling of wheelchair is to indirectly perceive the falling down of the user sitting on it. Heart rate and ECG are the primary physiological parameters for the wheelchair users' healthcare. WBANS contain sink node, heart rate sensor node and ECG sensor node to measure the heart rate of the users.

In addition to these sensors, a couple of sensors called pressure cushion and accelerator sensor are used to detect a human falling from the wheelchair and detect wheelchair falling down respectively. The aim of perceiving the surroundings is to monitor the emergency of wheelchair users. Environment parameters such as temperature, humidity and carbon monoxide are computed when the wheelchair shifts from one place to another. The smart phone is the core device of the people-centric networks, and is not only the gateway for transforming data and instruction between the nodes and internet, but also the interface between human and physical world. Wheelchairs, families or clinicians can remotely visit the data or control the nodes of WBSNs.

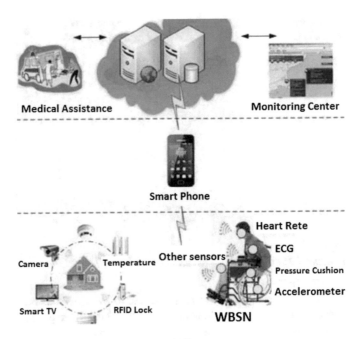

Fig. 13 Architecture of the healthcare system [11]

The data center platform connects different smart objects, builds the interoperation of the virtual nodes for smart objects and provide sharing of data to different applications via various interface. In concluding the above discussion it is obvious that the proposed method excels the other related methods in the sense that the former raises its portability and flexibility by providing a remote monitoring system.

2.7 Smart Home for Elderly and Disabled Persons

Home automation encompasses household activities such as centralized control of lighting, heating, ventilation and air conditioning appliances, security locks of gates and doors which would provide improved quality of life for the elderly and disabled persons who might otherwise have to oblige their caretakers. Commercial home automation systems are categorized as locally controlled systems and remotely controlled systems. The locally controlled systems use their automation system from within their home via a stationary or wireless interface. The remotely controlled systems use The IoT to allow the users to control their system from their mobile device, personal computer, or telephone.

Vishwajeet H. Bhide [12] has designed a fully smarthome environment monitoring various sensors. The key issue in designing a home automation system is to offer a cost-effective user-friendly interface on the host side, with the intention that the

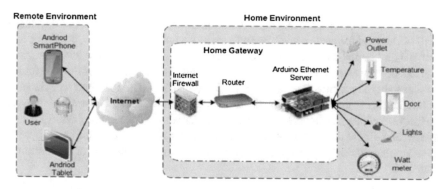

Fig. 14 Home automation [12]

devices can be easily monitored and organized. For this reason, cloud networking and data infrastructure are used to monitor, manage, and control their personal data points through the internet. The framework for the proposed work is shown in Fig. 14. The components include sensors, a home PC, a cloud server etc.

Different sensors for light, temperature, humidity etc. are used to gather the data to understand the environmental conditions and also to detect any fault in devices. The home PC is used to monitor sensor values, control the devices accordingly and send information about faulty devices to the cloud server. The cloud server on noticing the faults in the devices will send details to the owner. The aforesaid discussion shows that this novel technique would assist the elderly and disabled persons in monitoring and controlling the household activities without depending on their caretakers.

3 Conclusion and Future Research

This chapter presents a detailed survey on the applications of IoT. The key objective of the IoT is the "things" having different types of communication methods and belonging to different kinds of networks should be connected so as to share the required information.

In the future, first and foremost, the researchers have to focus on integrated IoT systems [13]. Moreover, if the number of the "things" in the network is significantly increased, proper protocols for the communication is required. Owing to the intro-duction of the IPv6 protocol, the IoT communications have been improved due to the large address space in it. Thus, the IPv6 over Low power Wireless Personal Area Networks (6LoWPAN) standard is likely to support the IoT communication in future. In view of the fact that different "things" are connected to the IoT network, security requirements are also varied according to network conditions. Since, the existing framework of security is not sufficient to keep the IoT secured, a safe and secured network is crucial for the IoT systems.

With the IoT being a product of information and communication technology, there will be demand for professionals with IT skills such as data scientists, user interface experts and digital-mechanical engineers. Also there will be a marked increment in the technology awareness of workers. Thus, instead of being afraid of the IoT, future workers should prepare themselves to work alongside it. The IoT apart from having robust cloud computing facilities, the smart Things in IoT should be easily deployed in the form of plug-n-play in any IoT framework. In the IoT, the data comes in smaller pieces and all the data is not needed. Hence the Big Data is required to collect only relevant data to filter irrelevant data and to handle massive amounts of unstructured data.

References

1. Luque, C.M.: Human can make the internet of things smarter. Harvard Business Review (2014). www.Hbr.org
2. Huang, T., Flanigan, J.: Human centered systems in the perspective of organizational and social informatics. J. UIUC **12**, 15–20 (1997)
3. Maguire, M.: Mehtods to support human-centred design. Int. J. Human Comput. Stud. **55**(4), 587–634 (2001)
4. Linjakumpu, T., Hartikainen, S., Klaukka, T., Veijola, J., Kivela, S.L., Isoaho, R.: Use of medications and polypharmacy are increasing among the elderly. J. Clin. Epidemiol. **55**(8), 809–817 (2002)
5. Jara, A.J., Zamora, M.A., Skarmeta, A.F.: Drug identification and interaction checker based on IoT to minimize adverse drug reactions and improve drug compliance. Pers. Ubiquitous Comput. **18**(1), 5–17 (2014)
6. Mulfari, D., Celesti, A., Fazio, M., Villari, M.: Human computer interface based on IoT embedded systems for users with disabilities. Internet of Things, User Centric IoT, pp. 376–383. Springer International Publishing (2014)
7. Wang, J., Cheng, Z., Zhang, M., Zhou, Y., Jing, L.: Design of a situation-aware system for abnormal activity detection of elderly people. In: Proceedings of the International Conference on Active Media Technology, pp. 561–571 (2012)
8. Sula, A., Spaho, E., Matsuo, K., Barolli, L., Xhafa, F., Miho, R.: An IoT based framework for supporting children with autism spectrum disorder. Information Technology Convergence, pp. 193–202. Springer Netherlands (2013)
9. Xhafa, F., Fernandez, R., Daradoumis, T., Barolli, L., Caballe, S.: Improvement of JXTA protocols for supporting reliable distributed applications in P2P systems. In: Proceedings of the International Conference on Network Based Information Systems, pp. 345–354. Springer, Berlin Heidelberg (2007)
10. Aishwarya, S.R., Ashish, R., Charitha, Prasanth, M.A., Savitha, S.C.: An IoT based accident prevention & tracking system for night drivers. Int. J. Innov. Res. Comput. Commun. Eng. **3**(4), 3493–3499 (2015)
11. Yang, L., Ge, Y., Li, W., Rao, W., Shen, W.: A home mobile healthcare system for wheelchair users. In: Proceedings of the IEEE 18th International Conference on Computer Supported Cooperative Work in Design, pp. 609–614 (2014)
12. Bhide, V.H.: A survey on the smart homes using internet of things. Int. J. Adv. Res. Comput. Sci. Manag. Stud. **2**(12), 243–246 (2014)
13. Kawamoto, Y., Nishiyama, H., Yoshimura, N., Yamamoto, S.: Internet of Things (IoT): present state and future prospects. IEICE Trans. Inf. Syst. **97**(10), 2568–2575 (2014)

A Survey on Internet of Things: Case Studies, Applications, and Future Directions

Kalavathy Perumal and Murali Manohar

Abstract Internet is recognized as the best innovative and influential establishment in the human history. The Internet of Things (IoT) represents the upcoming huge step in the Internet with its ability to gather, distribute, analyze and interpret data. The combination of Internet with RFID technologies, smart objects, sensor technologies, network components and electronics is IoT. IoT has been recognized as the promise of technology today and are well-defined as things belonging to the Internet aiding to the accessibility and supply of all of real-time data. Millions of devices are expected to be connected or networked into the IoT structure which requires massive dissemination of networks as well as the method of converting raw data into meaningful interpretations. The form of communication that is experienced now is either human-human or human-device, but Internet of Things (IoT) assures a promising type of communication, that is machine-machine (M2M). More influential smart phones, appliances, tablets and the applications that are similarly rich and powerful available for each, will enable buyers and business customers to interact seamlessly with companies altering the business processes. With businesses realizing that great digital experience to customers is no longer nice to have, but it is a make-or-break point for business has led to the integration and wider acceptance of Internet of Things. This chapter gives an idea of Internet of Things (IoT), the technologies that enable the implementation of IoT, the objectives and the future vision of IoT, the general layered architecture of IoT and the future growth challenges. This chapter also discusses the various applications of IoT and two case studies.

K. Perumal (✉) · M. Manohar
VIT Business School, VIT University, Vellore, Tamilnadu, India
e-mail: kala29.me@gmail.com

M. Manohar
e-mail: bmm@vit.ac.in

© Springer International Publishing AG 2017
D.P. Acharjya and M. Kalaiselvi Geetha (eds.), *Internet of Things:
Novel Advances and Envisioned Applications*, Studies in Big Data 25,
DOI 10.1007/978-3-319-53472-5_14

1 Introduction

With the numerous advancements in technology, the increasingly evolving technology is the system of inter connected objects, from books to cars, from electrical appliances to day today activities known as the 'Internet of things' (IoT). With the internet connecting people across the globe at an unprecedented pace and scale, the next thing is the interconnection between multiple objects thereby creating a smart environment. At present, IoT is used in numerous fields and is adopted steadily for the interchange of data between different entities. The data collected from the physical devices that are connected to these entities can be further used for forming new principles for human needs. The motivation for including IoT in our daily life is the accumulation of data from various sources and sharing it with millions of people around the world in a faster way.

The idea of the Internet of Things became standard through the Auto-ID center at MIT University and term was initially coined by Kevin Ashton in 1999. The RFID cluster defines IoT as the worldwide network of interconnected objects uniquely addressable based on standard communication protocols, incorporating ancient fields like Embedded Systems, management Systems and Automation, Wireless sensing element Networks to facilitate Device to Device (D2D) communication through the internet [1]. In 2003, we had Six billion people living on the planet and Five hundred million devices that were connected to the Internet. Cisco IBSG forecasts that there will be Twenty five billion devices connected to the Internet by 2015 and Fifty billion by 2020 which indicates that the connectivity of things around the globe will increase exponentially [2].

Internet of Things is progressing at a quick pace and is restructuring current static web into a completely integrated Future web [3]. Anyone can keep track of their belongings for many place and anytime with the network of devices interacting with one another and this will change the way people work, think and lead their lives. This technological phenomenon of Internet of Things combines innovative solutions and concepts of Information and Commutation systems that are associated with the following.

- *Ubiquitous Computing*: Ability of the networked objects to communicate anytime and anywhere.
- *Pervasive Computing*: Objects enhanced with processing power which means that the each individual object becomes a computer.
- *Ambient Intelligence*: Capability of the objects to capture and register changes in the physical environment thereby participating and interacting in the process.

IoT for this reason can also be defined as 'the ability of smart objects to communicate among each other and building networks of things' [4]. The technologies used in implementation of IoT are:

- *Radio Frequency Identification (RFID)*: It is used for automatic identification of anything they are attached acting as an electronic bar code and is in the design of microchips for wireless data communication.

- *Wireless Sensor Networks (WSN) Sensors*: The WSN network consists of hardware (consisting of power supply, processing units, sensor interfaces and transceiver units), communication stack (enables communication amongst the nodes), and middleware (platform independent sensor application development), secure data combination (for ensuring reliable data collected from sensors and for extending the lifetime of the network).
- *Addressing Schemes*: The basic for development of IoT is that the uniform resource network (URN)that makes replicas of resources accessed through address (Uniform resource Locator). The connectivity flow of internet from users (high level) to sensors (low level) that's available (through URN) and accessible (through URL).
- *Data Storage and Analytics*: The information created by IoT are to be stored and used for smart sensing using intelligence. To attain machine-driven decision making, storage solutions based on cloud are getting common. Temporal machine learning strategies supported by algorithms, neural networks and alternative computing techniques are necessary for machine-driven conclusions.
- *Visualization*: This plays a key role in attracting and understanding the IoT revolution to a user. This allows interaction of the user with the environment through images, diagrams or animations to communicate the message [5].

This chapter discusses the present developments in IoT analysis propelled by the requirement for convergence and applications in many knowledge domain technologies. Specifically, in Sect. 2, the general IoT design vision and challenges supported by the findings is discussed and in Sect. 3, many application domains in IoT with an approach in shaping them is discussed. Sections 4 and 5 presents case studies on few sensible environments are mentioned. Future vision of IoT is presented in Sect. 6. Finally the chapter concludes in Sect. 7.

2 Architecture of IoT

The general IoT architecture consists of the perception layer which is the Context aware tier, Middle ware layer, Network layer and Application layer. The context aware tier that perceives and procures information using the integrated hardware is also called as sensing layer. The RFID and WSNs form a part of perception layer.

The Middleware layer interposed between the application and network layer permits the developers to concentrate on the development method by concealment the hardware details. Along with measurability, interoperability and abstraction, this layer is accountable for providing services to the users. Besides effective delivery of services it additionally authenticates the users to a safer environment. The information Storage and Analytics and visualization form a of Middleware layer.

Aggregating the knowledge from numerous sources and routing it to correct destinations and also the secure data transfer over the sensing element networks is provided by the Network layer. The information transfers over the wireless network

Fig. 1 Layerwise functions
of wireless sensor network

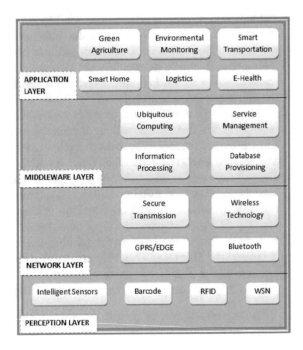

technology like Bluetooth, Infrared, 3G, Wi-Fi etc. is enabled by this layer. Data aggregation, addressing schemes form a part of the network layer.

The application layer is the uppermost layer of the IoT design. The delivery of all the services in varied fields of business like education, medical care, insurance, logistics, agriculture, media, environmental observance etc. is provided by this layer as shown in Fig. 1 [6].

2.1 Findings of IoT

The findings of IoT as compared to traditional information networks is to provide more extensive interconnection, more comprehensive intelligent service and more intensive information perception.

- *More extensive Interconnection*: It means the interconnection of all insightful or non-insightful physical objects like the interconnection between mobile devices and computers. The characteristics are extensiveness in the number of devices, the sort of devices and the mode of connectivity. The quantity of the connected devices can sharply rise from over hundred billions to over several billions (including a steep increase in the actuators, sensors and devices connected with RFID), the device communication and computation capability could also be greatly vary for the devices (some devices even might not have any process ability) and therefore the devices can be wired or wireless (the communication might be one hop or multiple hops).

- *More Comprehensive Intelligent Service*: The software design needs to be adaptive in order to provide intelligent services as the IoT environment is dynamic. Innovative software modeling techniques, delivery methods and mechanisms needs to be developed that can adapt to such changing IoT environment.
- *More Intensive Data Perception*: There could also be uncertainties in the data that is sensed from every sensing element within the following aspects: Non Uniformity, Inconsistency, Inaccuracy, Discontinuities, Incomprehensiveness and wholeness. It becomes troublesome to use the sensing element data directly, and there the challenge of efficient utilization arises thanks to unreliable sensory information [7].

2.2 Challenges

Internet of Things devices presents innate challenges as they are inhibited with low processing, memory, energy and communication capabilities.

- *Ubiquitous Deployment*: To confirm that the communication network is dependable and ascendible the key challenge is the assimilation of multi-technology networks in an exceedingly common all-IP network. IoT depends on the new IPv6 protocol to cover the measurability and addressing requirements, reliability and connectivity for its communication requirements, to eliminate this.
- *Interoperability and Standardization*: Normalization of IoT to produce effectual services IETF (The Internet Engineering Task Force) initiative on standardization to permit devices and objects to integrate with IPv6 and web services (better routing services, heightened reliability and security) will contribute vastly to the expansion of IoT. With IoT being open standard, IPv6 becomes a reliable protocol.
- *Security and Privacy*: To ensure privacy, security, integrity and confidentiality of the user data is another key challenge [8]. Built up on defined geographical boundary with numerous objects deployed, unauthorized persons access to objects needs to be prevented to ensure that the outlined practicality is not altered and also to ensure there is no physical damage [9].
- *Robustness in Connectivity*: IoT will need to work on providing reliable and enhanced connectivity in order to connect the humans and objects through sensors. Investment on energy garnering devices needs to be considered to improve connectivity with the assistance of energy mechanism.
- *Additional challenges are arising from the technological, economical, application and views*:

  - From the economic perspective, the economies of scale need to be provided in order to leverage the connected platform investment using new services that are supported by the existing modules.

- From an application perspective it's the massive information known as the big data [10]. From an IoT perspective of big data, it's mandatory to ensure that only solely relevant information is being extracted from the large databases [11].
- From the networking perspective, considering that different IoT applications present varied requirements in terms of bandwidth and potential it is important to ensure the end to end support for Quality of Service (QoS) [12].

3   Applications of Internet of Things

Emerging Internet of Things has impacted several applications these applications are classified based on scale, coverage, availability, repeatability, user experience, heterogeneity and impact. The categorized four different applications domains are (1) Personal and Home; (2) Enterprise; (3) Utilities; and (4) Mobile.

As depicted in Fig. 2 [6], at the home or individual level, it is the home or personal IoT, at the community level, it is the enterprise IoT, at the regional or national level it is the utility IoT and the one that is usually spread across every domain due to the scale of connectivity is the mobile IoT. Numerous IoT applications are already available and the upcoming applications of IoT promises to improve the quality of our lives. Few of these applications are:

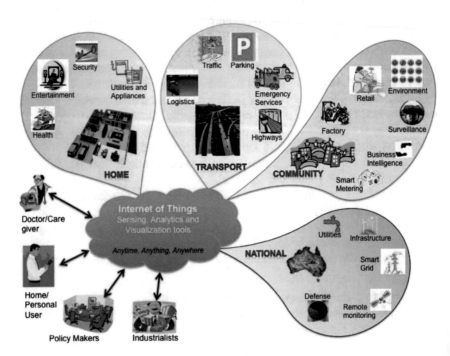

Fig. 2   Internet of things sensing analytics visualization tools

3.1 Smart Traffic System

Traffic problems are one of the major problems of the society. A prediction or fore-thought on traffic will improve the traffic situation and improve the traffic system. With an intelligent traffic monitoring system, a number of WSN and Sensor enabled communications are connected to provide sufficient data about the availability of the provision and logistics on the highway. This in turn will make the system smart and self-reliable. This intelligent traffic system can be called as Traffic IoT. This provides traffic information to be collected, processed, integrated and analyzed automatically and intelligently [13].

The model of smart traffic system has been applied already in many developed countries. A proposed methodology for intelligent transportation has been discussed in [14] for Jamaican transportation environment. Data mining that includes critical data, allows transport managers to make instantaneous decisions and the concerned authorities to methodically oversee a schedule events for a particular day. Logistics and be improved which can help in classifying peak times and off peak times, busy routes, as well as other irregularities at a more granulated level by using data mining. This enables the depot managers to make changes to deployments of the units based on the actual demand.

This smart traffic system eliminates the traditional traffic control systems that are unidirectional, huge implementation costs, dependency on the environmental conditions, etc. However, the limitations occur when the position of vehicle does not come in alignment the sensors and the sensor goes unresponsive, the sensor reading may not be accurate when the vehicle is under faulty condition.

3.2 Smart Environment

Technology with embedded smart sensors can oversee and communicate critical factors of the environment The common aspects of the environment are humidity, pressure, temperature etc. Some of the services that could be enabled based on a smart city concept are given below [15].

- *Monitoring Health of Buildings*: Required for proper maintenance of each building and the discovery of areas that are to be impacted with foreign agents.
- *Monitoring Quality of Air*: The quality of air can be monitored using IoT and also health applications can be connected the infrastructure.
- *Monitoring Noise*: The quantity of noise produced at any given hour can be measured by IoT and detection algorithms enable to identify the type of noise.
- *Smart Parking*: The vehicle users are directed to the best parking slot in the town with the help of smart displays and road sensors. The benefits include quicker time to find a parking slot which also includes less congestion of traffic, less vehicle emissions.

- *Smart Lighting*: The concentration of the street light can be adjusted depending on the weather conditions, presence of people and many other factors. The smart city infrastructure should include street lights for the service to work properly and can also include an error recognition system along with street light controllers.

The benefits of smart environment are ease of use, customization that could be done and the convenience it brings to our daily lives. The limitations include the cost and reliability.

3.3 Smart Home

Smart homes are of increasing popularity in the recent years due to the simplified and cheaper smart connectivity and home automation. With the boom of the IoT, the study and execution of home automation is gaining momentum. The benefits of smart homes are improved convenience, luxury, security and energy savings [16]. The commercially available smart home systems can be separated into two categories:

1. Locally controlled systems: Locally controlled systems via a stationary or wireless interface using the in-home controller allow control of the automation system within their home for the users.
2. Remotely controlled systems: Remotely controlled systems on the other hand can be controlled remotely from the personal computer of the user or via mobile device or via telephone. This is done with the use of an Internet connection or connecting with an already existing home security system [17].

The advantages of smart home include convenience and security; however the implementation cost will be a huge limitation.

3.4 Healthcare Monitoring Applications

Internet of Things application in Healthcare is being explored by major companies to improve the quality of care, improve the access to care and most significantly cut down on the charges of care [18]. The use of IoT in personalized Healthcare can be classified into Clinical healthcare and Remote Monitoring.

1. Clinical Care: It is the monitoring of hospitalized patients whose biological condition needs constant monitoring or attention. This enables automated flow of information in a continued basis and improved excellence of care which in turn removes the dependency of the caregiver to constantly engage in collection of data and investigation and lowers the cost of care.
2. Remote Monitoring: It is the possibility of providing monitoring to patients without visiting physicians. This is achieved by tiny, efficient wireless solutions connected via the IoT.

Some examples of applications of Internet of Things in healthcare include Monitoring an aging family member where elderly patients monitoring can be enabled using a sensor system that is waterproof and can be worn like a wristwatch. Scalable, continuous heart rate monitoring is a system that can monitor a patients: ECG Heart rate (including heart rate variability and dependability), activity level and respiration rate. Healthcare monitoring applications offers great benefits in personalized care, elderly care and women healthcare. The limitations include the security and privacy of sensitive data that is being shared across various participants.

3.5 Logistics and Supply Chain Management

RFID technologies with IoT provide numerous benefits to shop keepers as they can easily track the stocks and detect shoplifting and also can prevent out-of-stock of items by tracking constantly and placing an order automatically. With IoT, the logistics commodity has changed greatly to become an autonomous by a variety of applications available to support the needs [19].

One such application of Internet of Things in Supply chain management is the Smart Shopping Cart Subsystem. To assist the user in choosing goods, the smart shopping cart in a supermarket act as an assistant. Smart shopping cart ensures information flow is instant and interactive by interacting constantly with data center and smart shelves. The important functions of a smart shopping cart are as follows:

1. View: Here a buyer can get the particulars of the merchandises, including raw material supply data, the production data and shipping data and can also see the discounted commodity data, etc. at the same time.
2. Goods Location Map: This provides the user navigation information on the position information of goods so that the user can locate the good quickly.
3. Quick Settlement: Here user can add the merchandises to the cart (including virtual and physical cart), confirm and buy the merchandises after scanning using the RFID reader and then quickly complete the payment [20, 21].

4 Case Study on Multi Robot System

This case study is based on a multi-robot system where a robot accomplishes a stated mission without human intervention. In this proposed system, a robot provides information to another robot with information that will be needed to run live actions [22]. In this system, the challenge is on feeding every robot with environmental information for accuracy in action. Three techniques are used for the robot to get information from the environment. In the first technique, the information is given from an airplane that will regulate and achieve the robot mission. In the second technique, human control techniques are used for providing data. In the third technique, multimedia information is collected from the robots environment by a robot using cameras or sensors which in turn feeds the data to another robot.

For the case study, a mission against terrorism usually accomplished by a human team is considered. The current multi robot systems has lot of challenges on execution of the mission independently, wide range communication and scalability. The proposed multi robot system offers improved performance and establishes a sustainable solution that can be used for some serious missions, such as fighting against terrorism, war and violence (Fig. 3).

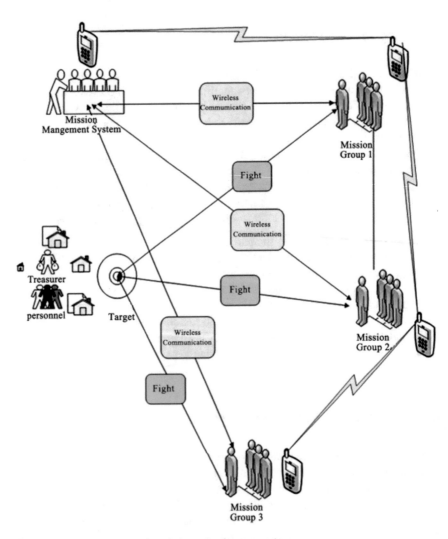

Fig. 3 Scenario of executing the mission using human combat team

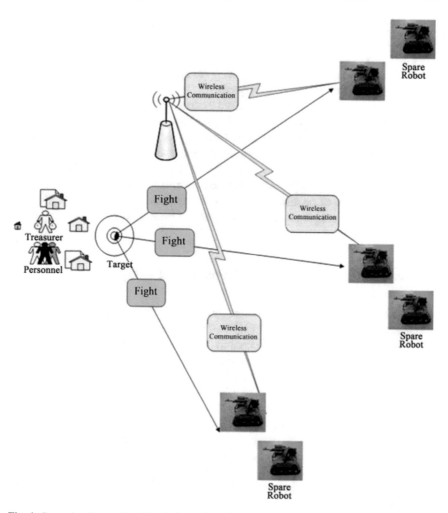

Fig. 4 Scenario of executing this mission using robots

In this proposed multi robot mission, each human is replaced by a robot and each target set for a soldier is executed by a robot. The sensing ability of a human is attained using sensors by the robot, the ability on vision is attained by cameras, the ability on speaking is attained by information communication between robots, the ability on thinking is attained by simulated intelligence techniques [23] and the ability on prediction is resolved by feedback algorithms [24]. The Fig. 4 shows the scenario of a mission execution by a human combat team where it is required to capture a terrorist either live or dead. The mission execution team has four well trained and well equipped soldiers who are provide with arms and wireless communication tools.

Below Fig. 4 shows the execution of this mission using robots. A completely armed robot that has sensors, camera, battery, a RFID tag, a wireless Internet connection card and the fighter tools, replaces each soldier. The target is fixed for the robot in the mission and can cooperate with rest of the robots. In case of movement of a terrorist from one area to another, an alarm is sent by the robot which is responsible for this region observation to the rest of the robots in mission advising them on the new state (position, style, and equipment) of the terrorist.

The various advantages of the proposed system include Scalability (where many number of robots can be connected using internet and also has the capability to improve the system with addition of new functionality to the web intelligent applications wither locally or globally), Least hardware complications (RFID tags or readers is the only hardware that might be needed), Multi applications suitability (For bigger missions, one robots applications can be collected to be one strong application), Communication over long distances (the media of robot communication is available any whereas the Internet connectivity is supported by satellite).

5 Case Study on Emergency Department in Hospitals

Applications of RFID in Hospitals An Emergency Department Case Study: In a hospital, the emergency department is the most busiest and difficult place. It is the center where patients (cases of accident injuries and/or acute sicknesses) need immediate and adequate care at the earliest time possible from both the physicians and the nurses. Preserving precise information in real time on the patients location is also a crucial issue for the medical workers as the patient leaves the premises with a registration record but not a medical record [24]. It is also important to reduce the patient wait time. This case study is about improving the safety of medical treatment through RFID tags in the emergency room for medical treatment process of the patients which includes an alerting mechanism on overly long queue and prolonged stay, tracking patients location thereby improving the quality of medical treatment, improved satisfaction of patients towards the hospital and allowing adequate time for treatment of emergency cases. The goal of the study is to plan a RFID query system with the major function to provide real-time data of the location of the patient with extremely long waiting time [25].

Figure 5 shows the procedure of most hospitals at emergency department where it starts with initial triage at the hospital entrance.

The relationship between the departments (radiology or laboratory) and the two way arrows depict the physician diagnosis as denoted in the above diagram. This indicates that such method is not essentially dependent on the decision of the physician. The five problem areas are highlighted in red which are long waiting time for consulting, unreasonably long stay for patients in the temporary observation area resulting in the patients leaving the hospital premises without an excuse. This in turn compels dissatisfaction toward the hospital and the physicians. Patients departure without payment, patients departure without excuse, no occupancy of sickbed makes patients go back to wait in the emergency center.

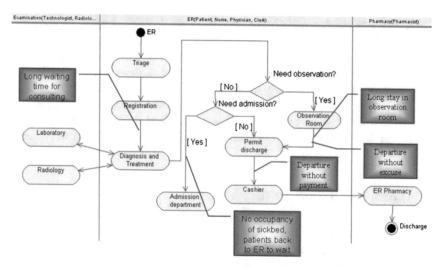

Fig. 5 The process at emergency department in most hospitals

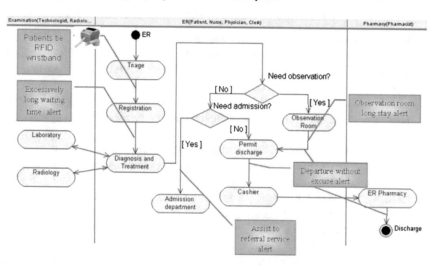

Fig. 6 The process at emergency department after RFID implementation

In Fig. 6, the application of RFID is used by the medical personnel in emergency department by instantly printing and placing a RFID wristband on the patients wrist at initial triage and is then is sent to the waiting area. In case the patient is not diagnosed by the physician in the time stipulated by the hospital, the management center is alerted with a short message where higher management may initiate an emergency protocol. Nevertheless, if the physician has made the patient wait for a predetermined amount of time though it is long, the RFID will not be read.

As shown in Fig. 6, the RFID system will automatically initiate short messages at the stages highlighted in Green due to which all the problem areas listed is addressed. The trial introduction of the RFID system at the hospital emergency center resolves the issues of extremely long waiting time of the patients, those who are not attended to by a medical personnel, unreasonably long waiting time of sickbed occupancy, excessively long provisional observation period and departure without excuse. This ensures that the hospital assures the five-step medical safety checklist (the right timing, the right patient, the right medication, the right dosage and the right approach).

6 Internet of Things Future Vision

The Internet of Things vision is still under development as there are many stakeholders in this development involved based on their interests and usage. Sensor based information gathering, organizing, mining and World Wide Web involves with the present vision, including the sensor based hardware. The three particular visions as per [6] are briefly discussed below. Figure 7 depicts the three main visions of IoT.

1. **Things Oriented Vision**: This vision is sustained by the point that anything can be tracked using RFID pervasive technologies and sensors the RFID functions with the basic philosophy that any object can be uniquely identified using the terms of Electronic Product Code (EPC). The future vision is dependent on the sensors and its ability to accomplish the things oriented vision. The data will be generated collectively with the assistanceof sensor type embedded system and sensors.

2. **Internet Oriented Vision**: This vision is supported by the necessity to create smart objects that are interconnected with the things having the same features of IP protocols. The sensor based objects are converted to a format that is understandable, distinctively recognized and the characteristics are constantly observed which forms the base for smart objects that are presumed to be microcomputers.

3. **Semantic Oriented Vision**: This vision is sustained by the point that numerous data will be at our disposal with the amount of sensors that collect the data. This huge amount of data needs to be administeredexpressively as the information could possibly be redundant. For better understanding and representation the non-manipulated data needs to be organized and processed in areasonable manner. The sets of data can be divided into homogeneous and heterogeneous formats which also mean that there are issues with interpreting the data with regards to interoperability which is reliant upon the semantic tools and techniques to process the data [26].

Fig. 7 The three main
visions of Internet of Things

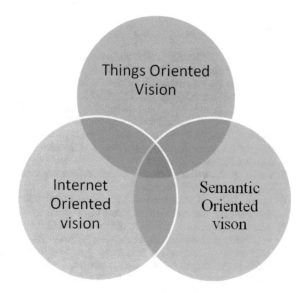

7 Conclusion

IoT targets to progressively spread the benefits of regular internet consistent connectivity, capability to remotely, data sharing, and so on. Just as Internet did, IoT will change the way of our life and world with lot of things connected. In this chapter we provided an introduction of Internet of Things, the evolution of connected devices and the various technologies that help in implementation of IoT. This chapter also discussed the objectives of IoT as compared to traditional information networks. The architecture of IoT was discussed that describes the various layers and the future vision that comprises the Internet Oriented Vision, the Things Oriented Vision and Semantic Oriented Vision. Various application areas are presented, providing guidance for future utilization of IoT concepts. The successful implementation of IoT is not without challenges, this chapter provided an overview of the challenges of implementing IoT and also discussed the other challenges at application, economical, and technological perspectives. The chapter also discussed on two case studies Multi Robot System and RFID Applications in Hospitals which explains the benefits of IoT technologies implementation. IoT has potential to drive integrated solutions that can make a difference allowing objects and people to be connected anyplace, anytime with anything and anyone. Though it is still early in the research cycle, the potential difference IoT can bring is very clear. There are ample opportunities for research and development to capitalize on the promise of IoT however the success of IoT mainstream will be based on the union of right standardization, improved sensors, cheaper and low power microprocessors, effective and efficient wireless protocols, the support of government, communities and established companies for the necessary development of IoT applications.

References

1. Kulkarni, A., Sathe, S.: Healthcare applications of internet of things: a review. Int. J. Comput. Sci. Inf. Technol. 5(5), 6229–6232 (2014)
2. Evans, D.: The internet of things: how the next evolution of the iternet is changing everything, cisco. Int. J. Internet 3(2), 123–132 (2011)
3. Wei, H.J., Shouyi, Y., Leibo, L., Zhen, Z., Shaojun, W.: A crop monitoring system based on wireless sensor network. Procedia Environ. Sci. 11(4), 558–565 (2011)
4. Dohr, A., Osprian, R.M., Drobics, M., Hayn, D., Schreier, G.: The internet of things for ambient assisted living. In: Proceedings of Seventh International Conference on Information Technology, pp. 345–349 (2010)
5. Pande, P., Padwalkar, A.R.: Internet of things a future of internet: a survey. Int. J. Adv. Res. Comput. Sci. Manag. Stud. 2(2), 354–361 (2014)
6. Gubbi, J., Buyya, R., Marusic, S., Palaniswami, M.: Internet of things: a vision, architectural elements, and future directions. Future Gener. Comput. Syst. 29(7), 1645–1660 (2013)
7. Khan, R., Khan, S.U., Zaheer, R., Khan, S.: Future internet: the internet of things architecture, possible applications and key challenges. In: Proceedings of the 10th International Conference on Frontiers of Information Technology, pp. 257–260 (2012)
8. Said, O., Masud, M.: A space unrestricted multi-robot combat internet of things system. Proc. Int. Conf. Adv. Internet Things 2, 56–62 (2012)
9. Matharu, G.S., Upadhyay, P., Chaudhary, L.: The internet of things: challenges and security issues. In: Proceedings of IEEE International Conference on Emerging Technologies, pp. 54–59 (2014)
10. Johnen, B., Scheele, C., Kuhlenkotter, B.: Learning robot behavior with artificial neural networks and a co-ordinate measuring machine. In: Proceedings of International Conference on Automation, Robotics and Applications, pp. 208–213 (2011)
11. Jara, A.J., Ladid, L., Skarmeta, A.: The internet of everything through IPv6: an analysis of challenges, solutions and opportunities. J. Wirel. Mob. Netw. Ubiquitous Comput. Dependable Appl. 4(3), 97–118 (2011)
12. Ma, H.D.: Internet of things: objectives and scientific challenges. J. Comput. Sci. Technol. 26(6), 919–924 (2011)
13. Singh, D., Tripathi, G.: A survey of internet-of-things: future vision, architecture, challenges and services. In: Proceedings of IEEE Conference on World Forum on Internet of Things, pp. 254–259 (2014)
14. Al-Sakran, H.O.: Intelligent traffic information system based on integration of internet of things and agent technology. Int. J. Adv. Comput. Sci. Appl. 6(2), 215–227 (2015)
15. Deans, C.: The design of an intelligent urban transportation system in jamaica based on the internet of things. In: Proceedings of the IEEE South East Conference, pp. 113–118 (2015)
16. Rajguru, S., Farooq, M.U.: Analysis of internet of things in a smart environment. Int. J. Enhanc. Res. Manag. Comput. Appl. 4(4), 40–48 (2015)
17. Vishwajeet, H., Bhide, K.: A survey on the smart homes using internet of things. Int. J. Adv. Res. Comput. Sci. Manag. Stud. 2(12), 243–246 (2014)
18. Bin, S., Yuan, L., Xiaoyi, W.: Research on data mining models for the internet of things. In: Proceedings of International Conference on Image Analysis and Signal Processing, pp. 127–132 (2010)
19. Niewolny, D.: How the internet of things is revolutionizing healthcare, freescale semiconductors. In: Proceedings of International Conference on Healthcare, pp. 211–219 (2013)
20. Farooq, M.U., Rajguru, S.: A review on internet of things. Int. J. Comput. Appl. 113(1), 113–119 (2015)
21. Zhu, J., Fu, B.: Research on supply chain simulation system based on internet of things. Proc. Int. Conf. Adv. Internet Things 5, 1–6 (2015)
22. Liu, D.T., Peng, Y., Peng, X.Y.: Online adaptive status prediction strategy for data-driven fault prognostics of complex systems. In: Proceedings of the International Conference on Prognostics and System Health Management, pp. 1–6 (2011)

23. Rao, R.Y.: Use PDCA management style to improve emergency patients healthcare. In: Proceedings of Annual Conference on Emergency Medicine, pp. 133–139 (2002)
24. Liaw, S.J., Hu, P., Liao, H.: Patients who leave emergency departments prematurely. J. Taiwan Emerg. Med. **4**(2), 40–49 (2002)
25. Huang, Y.C., Chu, C.P.: RFID applications in hospitals a case study for emergency department. J. Commun. Comput. **8**(1), 23–32 (2011)
26. Wei, H.J., Shouyi, Y., Leibo, L., Zhen, Z., Shaojun, W.: A crop monitoring system based on wireless sensor network. Procedia Environ. Sci. **11**(3), 558–565 (2011)

Internet of Things in Cloud Computing

Hemdan Ezz El-Din and D.H. Manjaiah

Abstract In the recent time, the advancement of network and communication systems developed new mechanisms for communicating and managing devices and it led to new area called Internet of Things (IoT). It means connecting devices through Internet for performing processes and services to provide prime services and needs of people in real life. So, it has become one of the most popular and important research directions for scientists and researchers to develop new approaches and procedures for performing communication process between devices in easy and efficient manner. To empower the connected devices with huge capabilities of computing and storage resources that can help to manage and handle billion of devices, there is a serious need in technology to support these capabilities. Cloud computing is one such essential technologies that can provide and enable IoT with the required capabilities and resources. This chapter reviews IoT in cloud computing systematically reviewed. Additionally, various research directions to enable cloud computing in IoT are presented.

1 Introduction

In the recent years, IoT has an imperative economic and societal impact for the future construction of communications to exchange information between people and things. The new regulation of future will be eventually, everything will be connected and intelligently controlled. The concept of IoT is becoming more pertinent to the realistic world due to the development of mobile devices, embedded and ubiquitous communication technologies, cloud computing and data analytics. Internet of Things (IoT) is made up of devices connected through the Internet to gather and collect information about the environment using sensors connected to things (i.e. devices).

H. Ezz El-Din (✉) · D.H. Manjaiah
Department of Computer Science, Mangalore University, Mangalore, India
e-mail: ezzvip@yahoo.com

D.H. Manjaiah
e-mail: manju@mangaloreuniversity.ac.in

© Springer International Publishing AG 2017
D.P. Acharjya and M. Kalaiselvi Geetha (eds.), *Internet of Things:
Novel Advances and Envisioned Applications*, Studies in Big Data 25,
DOI 10.1007/978-3-319-53472-5_15

These devices communicate and interact together to acquire, process and storage information in intelligent way.

In the last decades, cloud computing has started changing the ways people live and work. Cloud computing is a new paradigm that manage and arrange multi-tenancy, automated provisioning and usage accounting technologies through the Internet. Cloud computing now is more popular service that comes with more characteristics and advantages that change people life. The use of cloud computing in the IoT environment is an essential step to take the advantage of huge capabilities such as pool of available resources. The understanding of cloud environment will help researchers to design and develop new strategies and methods for IoT. These strategies and methods will enable the IoT to benefit from huge capabilities of the cloud such as processing and storage. Discussing the applications, challenges and future directions of the IoT is an important step in research community to introduce new opportunities for scientists and researchers in universities and research centres to help them in designing and developing new applications in various fields that will make life of people easier and comfortable in the future.

In the IoT environment, millions of devices are connected to each other which need to exchange information through the network (i.e. Internet) with huge capabilities such as high bandwidth, processing and storage capabilities. These huge capabilities can be provided by cloud computing. Researchers who are working in the Internet of Things (IoT) can use the cloud capabilities to design and develop applications that can create smart and intelligent environments such as smart homes and cities. Devices that are used in the IoT system generates huge amount of data which need to leverage cloud computing to scale cost effectively.

The reminder of this chapter is organized as follows: Sect. 2 provides foundations of cloud computing. Fundamentals of IoT is presented in Sect. 3. Section 4 provides challenges in cloud computing for enabling IoT and research directions to address the IoT challenges in real life. Finally, integration of IoT and cloud computing is discussed. Chapter conclusion is presented in Sect. 5.

2 Cloud Computing

Cloud computing is a new paradigm that manage and arrange multi-tenancy, auto-mated provisioning and usage accounting technologies through the Internet. Cloud computing now is more popular service that comes with more characteristics and advantages that change people life. The use of cloud computing in the IoT environment is an essential step to take the advantage of huge capabilities such as pool of available resources. The study of cloud computing is essential for understanding challenges and opportunities to enable IoT in the cloud. This section will discuss definition, characteristics and models of cloud computing technology. There are many definitions for cloud computing such as:

The National Institute of Standard and Technology (NIST) definition of cloud computing: Cloud computing is a model for enabling ubiquitous, convenient, on demand network access to a shared pool of configurable resources (e.g., networks, servers, storage, applications, and services) that can be rapidly provisioned and released with minimal management effort or service provider interaction [1].

The Cloud Security Alliance (CSA) definition of cloud computing: Cloud computing is an evolving term that describes the development of many existing technologies and approaches to computing into something different. Cloud separates application and information resources from underlying infrastructure, and the mechanisms used to deliver them. Cloud enhances collaboration, agility, scaling, and availability, and provides the potential for cost reduction through optimized and efficient computing [2].

The cloud computing has mainly five essential characteristics as follows [1]:

- *On-demand self-service*: Customers can automatically provision cloud capabilities without requiring human interaction for each request.
- *Broad network access*: Cloud capabilities are available via the network to access and use them by customers.
- *Resource pooling*: Huge pool of resources such as computing and storage are available to serve several customers in the same time.
- *Rapid elasticity*: Capabilities of cloud can be elastically provisioned and released in timely fashion manner for providing services to customers.
- *Measured service*: The cloud can automatically control and optimize resource used by leveraging a metering capability.

2.1 Cloud Computing Models

Cloud computing has two models; deployment and service models as shown in Fig. 1 as follows [1]:

1. **Deployment Models**: Deployment models in cloud computing are of four types as defined below.

   a. *Private Cloud*: Private cloud services are provided only for single organizations and not to the public use.
   b. *Public Cloud*: Public cloud services are exposed to the public use like individual users and organizations.

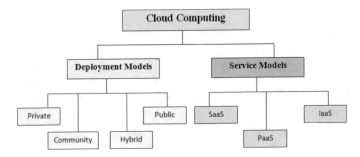

Fig. 1 Cloud computing models

    c. *Community Cloud*: Community cloud services are used and shared by many organizations to save costs, as compared to a private cloud.

    d. *Hybrid Cloud*: Hybrid cloud services can be combined among and distributed in multiple cloud types.

2. **Service Models**: Service models in cloud computing are of three types as defined below.

    a. *Software as a Service (SaaS)*: In SaaS, the user can able to use applications which are running on the cloud. These applications are accessible from several user devices. In this service model, the user cannot control the underlying cloud infrastructure like network, servers, operating systems and storage. There are many examples about SaaS services such as storage services.

    b. *Platform as a Service (PaaS)*: In PaaS, the user is able to deploy onto the cloud many types of applications using tools and programming languages supported by the provider. The user does not manage or control the underlying cloud infrastructure but has control over the deployed applications. Examples of PaaS are Windows Azure from Microsoft and Google App Engine from Google.

    c. *Infrastructure as a Service (IaaS)*: In IaaS, the user provision several resources such as processing, storage and networks. In addition to this, the user can develop and deploy to run arbitrary software. In this service model, the user does not manage or control the underlying cloud infrastructure but he can control over operating systems, storage, and deployed applications. An example of IaaS is Amazon EC2 service from Amazon.

3 Internet of Things

Internet of things will contribute to improve human life activity and life style by providing intelligent services like smart homes. This change in human life will help to create comfortable environment for people in their daily life. For example, when

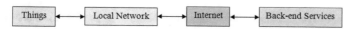

Fig. 2 The internet of things from an embedded systems point of view

an employee needs to finish his work report to submit his manager from the home, he will simply operate his computer which is connected to the Internet for finishing his work. IoT is a paradigm in which objects or humans are providing with a unique identifiers and it got an ability to transfer data automatically over the network without human interactions. The IoT gives us total control over anything which is at home or office from anywhere and at any time. In the IoT ecosystem, sensors devices connect through gateway to the Internet and then to cloud service provider. The end user can monitor and control remotely any environment through connecting to the cloud provider. This section introduces background and overview about IoT paradigm includes definition, classification, application and security in IoT.

Internet of Things means connecting devices to the Internet for performing processes and services that support our basics services and needs. The IoT can be considered as a novel paradigm that is rapidly gaining ground in the scenario of modern wireless networks and telecommunications. The Internet of Things (IoT) from an embedded systems point of view consists of four main components; hardware device (i.e. things), local network, communication network (i.e. the Internet) and finally, back-end services which are communicated to each other to exchange information as shown in Fig. 2.

3.1 Classifications of Internet of Things

Internet of Things can be classified into four types as shown in Fig. 3 such as: Internet of Nano Things, Internet of WiFi-enabled Things, Internet of Things for Smart Society, and Global-scaled Internet of Things [3].

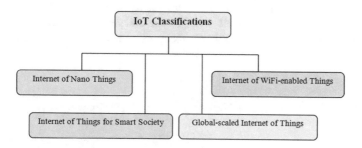

Fig. 3 Internet of things classifications

1. *Internet of Nano Things*: The Internet of Nano Things (IoNT) as a type of IoT that consist of nano-devices that are connected together through nanonetwork. In IoNT, it becomes possible to add a new dimension to the IoT by embedding nano-sensors to the various things and devices that surround us. The IoNT has many types in biological and multimedia fields which are called IoBNT and IoMNT respectively.

2. *Internet of WiFi-enabled Things*: Recently, WiFi has become an important type of wireless networks for wirelessly connecting various electronic devices to the Internet. When the WiFi enabled devices are connected together through the Internet that introduced new type of the IoT called Internet of WiFi-enabled Things that depends on the WiFi Connection.

3. *Internet of Things for Smart Society*: Nowadays, the concept of smart cities and world has become an attractive area for many researchers to develop new approaches and mechanisms for connecting objects in the society in an intelligent way to make an intelligent society (i.e. Smart World) by embedding sensors in all surrounding devices and objects to enable them to communicate and interact together in smart way by using new technologies like cloud computing. This brings a new type of IoT called internet of things for smart society.

4. *Global-scaled Internet of Things*: Global-scaled internet of things is a type of IoT that utilized in global-scaled area such as satellite system and Unmanned Aerial Vehicle (UAV). These systems remotely connect to various devices in ground which are connected to sensors to sense and acquire data in easy and efficient manner. An example about global-scaled internet of things is Tsunami Detection System.

3.2   Applications of Internet of Things

Internet of Things has become an important paradigm for providing smart and intelligent applications in several fields such as military, healthcare, transportation, agriculture, logistics, smart society, energy, security and emergencies. These applications will improve and change the quality of our lives to better level of life. The IoT has various applications such as; smart society, industrial control system and smart manufacturing, smart agriculture, healthcare ecosystem and trade and logistics as shown in Fig. 4 [4–8].

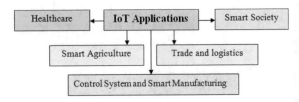

Fig. 4 Internet of things applications

1. *Smart Society*: Smart society is the concept of making smart world which makes people life easy and comfortable. The advanced development in smart technologies enables the IoT to effect in changing people life style. There are many applications in smart society domain such as home automation and smart cities as follows:

   - *Home Automation*: The IoT can be used in home automation to provide services as; monitoring of energy and power consumption to save cost and resources, remotely control in operating appliances to avoid accidents and intrusion detection and prevention to protect homes from stealing.
   - *Smart Cities*: The IoT can be used in smart cities to provide services as; intelligent highways with warning messages for an unexpected events like accidents or traffic jams, monitoring of vehicles and pedestrian levels to optimize driving and walking routes, monitoring of parking spaces availability inside the city.

2. *Industrial Control System and Smart Manufacturing*: Internet of things can be used in industrial control systems to improve their performance by making them more smart with taking in consideration necessary factors like safety and availability factors to guarantee continues in business and save people life. Industrial control system can use the IoT for many purpose as; auto-diagnosis of machines in control system, monitoring of toxic gas and oxygen levels inside chemical plants to ensure workers and goods safety, control of temperature inside industrial and medical fridges with sensitive merchandise and monitoring of ozone levels during the drying meat process in factories of food manufacturing.

3. *Smart Agriculture*: Agriculture is an important field that provide people and society with food so that there are serious needs to improve and enhance an agriculture system by using smart and intelligent technologies that are introduced by using the IoT. The IoT will improve operational efficiency and productivity in agriculture system. There are many benefits from IoT in agriculture field such as; monitoring soil moisture and trunk diameter in vineyards to control the amount of sugar in grapes and grapevine health, control micro-climate conditions to maximize the production of fruits and vegetables and its quality, location and identification of animals grazing in open pastures or location in big stables, prevent fungus and other microbial contaminants, control the exact conditions of plants grown in water to get the highest efficiency crops, control of growing conditions of the offspring in animal farms to ensure its survival and health and study of ventilation and air quality in farms and detection of harmful gases from excrements.

4. *Healthcare and Medical*: There are many services and benefits of using the IoT in healthcare and medical field like; remote monitoring of patients, tracking of drugs, identification and authentication of people. The IoT also can use in healthcare as assistance for elderly or disabled people living independent, control of conditions inside freezers storing vaccines, medicines and organic elements, monitoring of conditions of patients inside hospitals and measurement of Ultraviolet sun rays that warn people not to be exposed in certain hours.

5. *Trade and Logistics*: Advanced development in the IoT paradigm help to develop and manage shopping of products from online websites in addition to using IoT

in trade and logistics by embedding tags and sensors in roads and products for monitoring and tracking them and.

- *Trade*: IoT can be used in trade for many purposes such as; monitoring of storage conditions, product tracking, payment processing based in location or activity duration for public transport, getting advices in the point of sale according to customer habits, preferences, presence of allergic components for them or expiring dates and control of rotation of products in shelves and warehouses to automate restocking processes.
- *Logistics*: In logistics, IoT can be used for monitoring of vibrations, strokes, container openings for insurance purposes, Search of individual items in big surfaces like warehouse and warning emission on containers storing inflammable goods closed to others containing explosive material.

3.3   Security and Challenges in Internet of Things

Security issues for enabling IoT for various applications that are important to allow users to use IoT services in easy and secure environment because the devices communicate and interact through the Internet which make them easy to attack by malicious users. They are exploiting vulnerabilities and weakness in communication system. In the recent time, security of IoT became an attractive area for many researchers to develop mechanisms for handling security issues of IoT system. There are many security issues should be addressed in IoT. These are security and privacy of users in the IoT environment, authentication and authorization for IoT users, secure communication system between devices in IoT system, strong registration system for IoT users, and availability and safety of IoT infrastructure in industrial applications.

The Internet of things has many challenges. Few of these challenges are integration of IoT with the current existing communication systems, development of security mechanisms for secure communication, addressing and discovery of things in IoT network, manage and analyze large amount of structure and unstructured, semi-structure data that generated from IoT things, security and privacy in IoT paradigm, IoT standardization includes interoperability, radio access level issues, and semantic interoperability, embedding sensors for complex environment, connecting end-to-end and enabling applications, and above all power and energy consumption.

3.4   Future Research Directions in Internet of Things

There are several future research directions for researchers and scientists working in internet of things [9]. Few of these directions are listed below.

1. Develop new techniques and methods to secure IoT infrastructure.
2. Propose methodologies for performing digital forensics in IoT environment.

3. Provide new standards and protocol to enable IoT.
4. Efficient energy consumption algorithms and mechanisms.
5. Handle and manage large amount of data collected from sensors.
6. Transport protocols to use in IoT scenarios.
7. Privacy mechanism to protect user information.
8. Authentication system to authenticate access to IoT infrastructures.
9. Map a reference to a description of a specific object and the related identifier using Object Name Servers (ONS).

3.5 Practical System in Internet of Things

This section, a real IoT application called Tsunami Detection System is discussed. The Tsunami detection system [3] is introduced by Yuichi Kawamoto et al. (2014). This system is used to facilitate early detection of the tsunami using real-time observation of the sea level in Japan. Large number of buoys equipped with sensors and small earth stations are planned to be deployed around Japan in this system. The sensors are used to measure the fluctuation of the wave height while the small earth stations send the data gathered from the sensors to the satellite. To cover whole Japan to detect a tsunami as early as possible, it is necessary to deploy numerous sensor terminals. In the case where the sensor terminals are deployed in a large scale, it is difficult to send the data gathered from the sensors directly to the base station

Fig. 5 Tsunami detection system

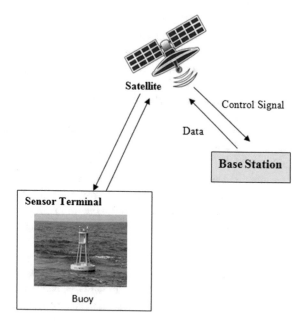

on the ground by ground-based wireless networks due to the remotely located base station. Therefore, in this system, the data is sent from the sensor terminals to the base station via a satellite. Since the satellite has a large coverage, it is possible to collect data from sensor terminals deployed throughout Japan. The construction of Tsunami detection system is shown in Fig. 5 which consists from many components such as sensors, small earth stations, base stations, and satellites to construct the detection system. In this system; the sensors are used to measure the change of the Z-axis position by using the GPS. The change of the Z-axis position implies the change of the sea level. The collected data is sent to the base station via a satellite and then the data are analyzed to distinguish the tsunami from normal waves. Thus, the occurrence of a tsunami can be detected at the base station. This system will help early to detect of tsunami before it occurs for protect people life.

4   Internet of Things in the Cloud

Internet of things with cloud computing led to a new trend and paradigm which provide an efficient managing and processing of data generated from sensors using the cloud. Cloud computing is an interesting area for researchers to enable the IoT by utilizing the huge capabilities of cloud such as high processing and storage. In the IoT environment, sensors can generate huge amount of data that need special processing and storage capabilities that can be provided by the cloud service providers. The cloud service providers can provide on demand, fast, scalable, efficient, and flexible data center to enable the IoT services.

Integration of IoT with cloud computing which are two different technologies produced a new paradigm called CloudIoT. This integration will help to perform efficient services in various types of applications like military, healthcare, agriculture, home automation and smart cities. The process of integration IoT in cloud need to satisfy some issues such as: combine services offered by multiple vendors and be scale to support large number of customers in a distributed and reliable way. Several advantages can be derived from the integration of cloud and IoT; Firstly, IoT can benefit from the huge and unlimited capabilities and resources of cloud computing. Secondly, Cloud can benefit from the integration with the IoT by extending its scope to deal with devices in real world that are distributed in different locations in addition to deliver new services for a lot of people in real world. The CloudIoT paradigm will enable new and novel scenarios for providing intelligent services and applications [4] based on the use of cloud through the things. A conceptual CloudIoT framework that integrate cloud services with user applications and things is used to provide the services of internet of things for end customers and users as shown in Fig. 6. The conceptual framework is used to illustrate the concept of integration both of cloud and IoT to provide best services in timely fashion manner for improving daily life of people. The CloudIoT framework consists from three layers; first layer to provide a network of things, then second layer provide cloud computing services and

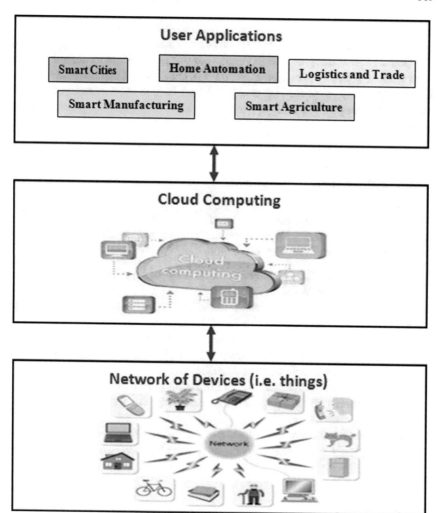

Fig. 6 Conceptual CloudIoT framework

capabilities to enable IoT applications and the third layer provide IoT applications and services for end users.

4.1 Challenges to Enable Internet of Things in Cloud Computing

Cloud computing can be used to enable the IoT services but there are many challenges in the integration of cloud computing and IoT [10, 11]. These are supporting application elasticity to satisfy the highly dynamic resource demand in the IoT environment, reliability in the cloud to support the IoT application availability, security and privacy for securing data of the users who use IoT services through the cloud, management of energy to consume the power of applications, provide interoperability and portability of the cloud to enable IoT user for access an interoperable cloud ecosystem, managing and analyzing large amount of data which are generated from sensors.

4.2 Future Research Directions

There are many research directions according to the challenges mentioned above that are opened for researchers in the IoT domain to handle and integrate with cloud computing [12]. Few of them are listed below.

1. *Elasticity of Application Services*: Elasticity of application services and autonomous adaptation of services.
2. *Scalability of Cloud Infrastructure*: To manage and scale of processing, storage and networks across multiple distributed locations.
3. *Security and Privacy*: Secure multi-tenant and multi-users environments and support isolation of services and applications for each user, and Integrity and privacy techniques to secure data in cloud storages.
4. *Energy Efficient Management*: Better utilization of resources by supporting and providing energy efficiency models load balancing and optimization algorithms for cloud infrastructures.
5. *Cloud Interoperability and Portability*: Standards for support integration of cloud computing and IoT and virtual appliances portability for cloud service providers.

5 Conclusion

Recently, Internet of Things (IoT) has potential significantly in several fields such as business, healthcare, energy management, agriculture, manufacturing, and transportation. The IoT is made up of appliances connected through the Internet that is capable of gathering information about surrounding environment to make appropriate action by influencing the environment autonomously. Cloud Computing is an emerging revolutionary technology that has started changing the ways people live and work. Cloud computing provide huge processing and storage capabilities that can be used by the IoT industry for performing high availability and reliability in providing services for people in the real world. This chapter is written to provide researchers an overview and background to understand cloud computing infrastructure, Internet of Thing (IoT) technology in addition to how integrate cloud computing with IoT for providing efficient and effective services for users.

References

1. Mell, P., Grance, T.: The nist definition of cloud computing. White Paper, National Institute of Standards and Technology, pp. 1–7 (2011)
2. Cloud Security Alliance: Security guidance for critical areas of focus in cloud computing V 2.1. San Francisco, California, pp. 1–76 (2009)
3. Kawamoto, Y., Nishiyama, H., Kato, N., Yoshimura, N., Yamamoto, S.: Internet of Things (IoT): present state and future prospects. IEICE Trans. Inf. Syst. **E97**(10), 2568–2575 (2014)
4. Gubbi, J., Buyya, R., Marusic, S., Palaniswamia, M.: Internet of Things (IoT): a vision, architectural elements, and future directions. Future Gener. Comput. Syst. **29**(7), 1645–1660 (2013)
5. Atzori, L., Iera, T., Morabito, G.: The internet of things: a survey. Comput. Netw. **54**(15), 2787–2805 (2010)
6. Li, S., Xu, L.D., Zhao, S.: The internet of things: a survey. Inf. Syst. Front. **17**(2), 243–259 (2014)
7. Andrew, W., Agarwal, A., Xu, L.D.: The internet of things a survey of topics and trends. Inf. Syst. Front. **17**(2), 261–274 (2014)
8. Farooq, M.U., Waseem, M., Mazhar, S., Khairi, A., Kamal, T.: A review on internet of things (IoT). Int. J. Comput. Appl. **113**(1), 1–7 (2015)
9. Bhavana, A.: Evaluating perception, characteristics and research directions for internet of things (IoT): an investigational survey. Int. J. Comput. Appl. **121**(4), 13–19 (2015)
10. Llorente, I.M.: Key challenges in cloud computing to enable future internet of things. In: Proceedings of 4th EU-Japan Symposium on New Generation Networks and Future Internet, pp. 8–11 (2012)
11. Suna, E., Zhang, X., Li, Z.: The internet of things (IOT) and cloud computing (CC) based tailings dam monitoring and pre-alarm system in mines. Saf. Sci. **50**(4), 811–815 (2012)
12. Botta, A., Donato, W.D., Persico, V., Pescape, A.: On the integration of cloud computing and internet of things. In: Proceedings of International Conference on Future Internet of Things and Cloud, pp. 23–30 (2014)

|‖| || ‖|‖ |‖|‖‖‖‖‖| |‖|‖ |‖| ‖ |‖|‖ ‖ |‖‖‖|‖‖ |‖‖ ‖|

Printed in the United States
By Bookmasters